Algorithms and Computation in Mathematics • Volume 3

Editors

Manuel Bronstein   Arjeh M. Cohen
Henri Cohen   David Eisenbud
Bernd Sturmfels

Springer

Berlin
Heidelberg
New York
Hong Kong
London
Milan
Paris
Tokyo

Neal Koblitz

Algebraic Aspects of Cryptography

With an Appendix on
Hyperelliptic Curves by
Alfred J. Menezes,
Yi-Hong Wu, and
Robert J. Zuccherato

With 7 Figures

Springer

Neal Koblitz

Department of Mathematics
University of Washington
Seattle, WA 98195, USA
e-mail:
koblitz@math.washington.edu

Yi-Hong Wu

Department of Discrete and
Statistical Sciences
Auburn University
Auburn, AL 36849, USA

Alfred J. Menezes

Department of Combinatrics
and Optimization
University of Waterloo
Waterloo, Ontario
Canada N2L3G1
e-mail:
ajmeneze@math.uwaterloo.ca

Robert J. Zuccherato

Entrust Technologies
750 Heron Road
Ottawa, Ontario
Canada K1V1A7
e-mail:
robert.zuccherato@entrust.com

1st ed. 1998. Corr. 2nd printing 1999, 3rd printing 2004

Mathematics Subject Classification (2000): 11T71, 94A60, 68P25, 11Y16, 11Y40

Cataloging-in-Publication Data applied for

A catalog record for this book is available from the Library of Congress.

Bibliographic information published by Die Deutsche Bibliothek
Die Deutsche Bibliothek lists this publication in the Deutsche Nationalbibliografie;
detailed bibliographic data is available in the Internet at http://dnb.ddb.de

ISSN 1431-1550

ISBN 978-3-642-08332-7

Springer-Verlag is a part of Springer Science+Business Media

springeronline.com

© Springer-Verlag Berlin Heidelberg 2010
Printed in Germany

Cover design: *design & production* GmbH, Heidelberg

Printed on acid-free paper   46/3142db - 5 4 3 2 -

Preface

This book is intended as a text for a course on cryptography with emphasis on algebraic methods. It is written so as to be accessible to graduate or advanced undergraduate students, as well as to scientists in other fields. The first three chapters form a self-contained introduction to basic concepts and techniques. Here my approach is intuitive and informal. For example, the treatment of computational complexity in Chapter 2, while lacking formalistic rigor, emphasizes the aspects of the subject that are most important in cryptography.

Chapters 4–6 and the Appendix contain material that for the most part has not previously appeared in textbook form. A novel feature is the inclusion of three types of cryptography – "hidden monomial" systems, combinatorial–algebraic systems, and hyperelliptic systems – that are at an early stage of development. It is too soon to know which, if any, of these cryptosystems will ultimately be of practical use. But in the rapidly growing field of cryptography it is worthwhile to continually explore new one-way constructions coming from different areas of mathematics. Perhaps some of the readers will contribute to the research that still needs to be done.

This book is designed not as a comprehensive reference work, but rather as a selective textbook. The many exercises (with answers at the back of the book) make it suitable for use in a math or computer science course or in a program of independent study.

I wish to thank the participants in the Mathematical Sciences Research Institute's Summer Graduate Student Program in Algebraic Aspects of Cryptography (Berkeley, 16–27 June 1997) for giving me the opportunity to test-teach parts of the manuscript of this book and for finding errors and unclarities that needed fixing. I am especially grateful to Alfred Menezes for carefully reading the manuscript and making many valuable corrections and suggestions. Finally, I would like to thank Jacques Patarin for letting me report on his work (some of it not yet published) in Chapter 4; and Alfred Menezes, Yi-Hong Wu, and Robert Zuccherato for agreeing to let me include their elementary treatment of hyperelliptic curves as an Appendix.

Seattle, September 1997   *Neal Koblitz*

Contents

Chapter 1. Cryptography

Broadly speaking, the term *cryptography* refers to a wide range of security issues in the transmission and safeguarding of information. Most of the applications of algebra and number theory have arisen since 1976 as a result of the development of *public key* cryptography.

Except for a brief discussion of the history of *private key* cryptography (pre-1976), we shall devote most of this chapter to the (generally more interesting) questions that arise in the study of public key cryptosystems. After discussing the idea of public key cryptography and its importance, we next describe certain prototypical public key constructions.

§ 1. Early History

A cryptosystem for message transmission means a map from units of ordinary text called *plaintext message units* (each consisting of a letter or block of letters) to units of coded text called *ciphertext message units*. The idea of using arithmetic operations to construct such a map goes back at least to the Romans. In modern terminology, they used the operation of addition modulo N, where N is the number of letters in the alphabet, which we suppose has been put in one-to-one correspondence with $\mathbb{Z}/N\mathbb{Z}$. For example, if $N = 26$ (that is, messages are in the usual Latin alphabet, with no additional symbols for punctuation, numerals, capital letters, etc.), the Romans might encipher single letter message units according to the formula $C \equiv P + 3 \pmod{26}$. This means that we replace each plaintext letter by the letter three positions farther down the alphabet (with the convention that $X \mapsto A$, $Y \mapsto B$, $Z \mapsto C$). It is not hard to see that the Roman system – or in fact any cryptosystem based on a permutation of single letter message units – is easy to break.

In the 16th century, the French cryptographer Vigenère invented a variant on the Roman system that is not quite so easy to break. He took a message unit to be a block of k letters – in modern terminology, a k-vector over $\mathbb{Z}/N\mathbb{Z}$. He then shifted each block by a "code word" of length k; in other words, his map from plaintext to ciphertext message units was translation of $(\mathbb{Z}/N\mathbb{Z})^k$ by a fixed vector.

Much later, Hill [1931] noted that the map from $(\mathbb{Z}/N\mathbb{Z})^k$ to $(\mathbb{Z}/N\mathbb{Z})^k$ given by an invertible matrix with entries in $\mathbb{Z}/N\mathbb{Z}$ would be more likely to be secure than Vigenère's simple translation map. Here "secure" means that one cannot easily figure out the map knowing only the ciphertext. (The Vigenère cipher, on the other hand, can easily be broken if one has a long string of ciphertext, by analyzing the frequency of occurrence of the letters in each arithmetic progression with difference k. It should be noted that, even though the Hill system cannot be easily broken by frequency analysis, it is easy to break using linear algebra modulo N if you know or can guess a few plaintext/ciphertext pairs.)

For the most part, until about 20 years ago only rather elementary algebra and number theory were used in cryptography. A possible exception was the use of shift register sequences (see [Golomb 1982] and Chapter 6 and §9.2 of [Lidl and Niederreiter 1986]).

Perhaps the most sophisticated mathematical result in cryptography before the 1970's was the famous theorem of information theory [Shannon 1949] that said, roughly speaking, that the only way to obtain perfect secrecy is to use a *one-time pad*. (A "one-time pad" is a Vigenère cipher with period $k = \infty$.)

The first harbinger of a new type of cryptography seems to have been a passage in a book about time-sharing systems that was published in 1968 [Wilkes 1968, p. 91-92]. In it, the author describes a new *one-way cipher* used by R. M. Needham in order to make it possible for a computer to verify passwords without storing information that could be used by an intruder to impersonate a legitimate user.

In Needham's system, when the user first sets his password, or whenever he changes it, it is immediately subjected to the enciphering process, and it is the enciphered form that is stored in the computer. Whenever the password is typed in response to a demand from the supervisor for the user's identity to be established, it is again enciphered and the result compared with the stored version. It would be of no immediate use to a would-be malefactor to obtain a copy of the list of enciphered passwords, since he would have to decipher them before he could use them. For this purpose, he would need access to a computer and even if full details of the enciphering algorithm were available, the deciphering process would take a long time.

In [Purdy 1974] the first detailed description of such a *one-way function* was published. The original passwords and their enciphered forms are regarded as integers modulo a large prime p, and the "one-way" map from $\mathbb{Z}/p\mathbb{Z}$ to $\mathbb{Z}/p\mathbb{Z}$ is given by a polynomial $f(x)$ which is not hard to evaluate by computer but which takes an unreasonably long time to invert. Purdy used $p = 2^{64} - 59$ and $f(x) = x^{2^{24}+17} + a_1 x^{2^{24}+3} + a_2 x^3 + a_3 x^2 + a_4 x + a_5$, where the coefficients a_i were arbitrary 19-digit integers.

§ 2. The Idea of Public Key Cryptography

Until the late 1970's, all cryptographic message transmission was by what can be called *private key*. This means that someone who has enough information to encrypt messages automatically has enough information to decipher messages as well. As a result, any two users of the system who want to communicate secretly must have exchanged keys in a safe way, e.g., using a trusted courier.

The face of cryptography was radically altered when Diffie and Hellman invented an entirely new type of cryptography, called *public key* [Diffie and Hellman 1976]. At the heart of this concept is the idea of using a one-way function for encryption.

Definition 2.1. Speaking informally, we say that a one-to-one function $f : X \to Y$ is "one-way" if it is easy to compute $f(x)$ for any $x \in X$ but hard to compute $f^{-1}(y)$ for most randomly selected y in the range of f.*

The functions used for encryption belong to a special class of one-way functions that remain one-way only if some information (the "decryption key") is kept secret. Again using informal terminology, we can define a *public key encryption function* (also called a "trapdoor" function) as a map from plaintext message units to ciphertext message units that can be feasibly computed by anyone having the so-called "public" key but whose inverse function (which deciphers the ciphertext message units) cannot be computed in a reasonable amount of time without some additional information (the "private" key).

This means that everyone can send a message to a given user using the same enciphering key, which they simply look up in a public directory. There is no need for the sender to have made any secret arrangement with the recipient; indeed, the recipient need never have had any prior contact with the sender at all.

It was the invention of public key cryptography that led to a dramatic expansion of the role of algebra and number theory in cryptography. The reason is that this type of mathematics seems to provide the best source of one-way functions. Later we shall discuss the most important examples.

A curious historical question is why public key cryptography had to wait until 1976 to be invented. Nothing involved in the idea of public key cryptography or the early public key cryptosystems required the use of 20th century mathematics. The first public key cryptosystem to be used in the real world – the RSA system (see below) – uses number theory that was well understood by Euler. Why had it not occurred to Euler to invent RSA and offer it to the military advisers of Catherine the Great in gratitude for her generous support for the Russian Imperial Academy of Sciences, of which he was a member?

A possible reason for the late development of the concept of public key is that until the 1970's cryptography was used mainly for military and diplomatic purposes, for which private key cryptography was well suited. However, with the increased computerization of economic life, new needs for cryptography arose. To cite just one obvious example, when large sums of money are transferred electronically, one must be able to prevent white-collar thieves from stealing funds and

---

* In some situations one wants a one-way function to have a stronger property, namely, that it is hard to compute any partial information about $f^{-1}(y)$ (for instance, whether it is an odd or even number) for most randomly selected y.

nosy computer hackers (or business competitors) from monitoring what others are doing with their money. Another example of a relatively new use for cryptography is to protect the privacy of data (medical records, credit ratings, etc.). Unlike in the military or diplomatic situation – with rigid hierarchies, long-term lists of authorized users, and systems of couriers – in the applications to business transactions and data privacy one encounters a much larger and more fluid structure of cryptography users. Thus, perhaps public key cryptography was not invented earlier simply because there was no real need for it until quite recently.

Another reason why RSA was not likely to have been discovered in Euler's time is that in those days all computations had to be done by hand. To achieve an acceptable level of security using RSA, it would have been necessary to work with rather large integers, for which computations would have been cumbersome. So Euler would have had difficulty selling the merits of RSA to a committee of skeptical tsarist generals.

In practice, the great value of public key cryptography today is intimately connected with the proliferation of powerful computer technology.

2.1 Tasks for Public Key Cryptography

The most common purposes for which public key cryptography has been applied are:

(1) *confidential message transmission*;

(2) *authentication* (verification that the message was sent by the person claimed and that it hasn't been tampered with), often using *hash functions* (see §3.2) and *digital signatures* (see §3.3); *password* and *identification* systems (proving authorization to have access to data or a facility, or proving that you are who you claim to be); *non-repudiation* (guarding against people claiming not to have agreed to something that they really agreed to);

(3) *key exchange*, where two people using the open airwaves want to agree upon a secret key for use in some private key cryptosystem;

(4) *coin flip* (also called *bit commitment*); for example, two chess players in different cities want to determine by telephone (or e-mail) who plays white;

(5) *secret sharing*, where some secret information (such as the password to launch a missile) must be available to k subordinates working together but not to $k - 1$ of them;

(6) *zero knowledge proof*, where you want to convince someone that you have successfully solved a number-theoretic or combinatorial problem (for example, you have found the square root of an integer modulo a large unfactored integer, or you have 3-colored a map) without conveying any knowledge whatsoever of what the solution is.

These tasks are performed through various types of *protocols*. The word "protocol" simply means an orderly procedure in which people send messages to one another.

In §§3–5 we shall describe several usable cryptosystems that perform one or more of the above tasks. We should caution the reader that the cryptosystems described in this book are *primitives*. In cryptography the term "primitive" means a basic ingredient in a cryptosystem. In order to construct a practical system one generally has to modify and combine these primitives in a careful way so as to simultaneously achieve various objectives related to security and efficiency. For the most part we shall not deal with the practical issues that arise when one does this. The best general reference for such issues is the *Handbook of Applied Cryptography* [Menezes, van Oorschot, and Vanstone 1996].

2.2 Probabilistic Encryption

Most of the number theory based cryptosystems for message transmission are *deterministic*, in the sense that a given plaintext will always be encrypted into the same ciphertext by anyone. However, deterministic encryption has two disadvantages: (1) if an eavesdropper knows that the plaintext message belongs to a small set (for example, the message is either "yes" or "no"), then she can simply encrypt all possibilities in order to determine which is the supposedly secret message; and (2) it seems to be very difficult to *prove* anything about the security of a system if the encryption is deterministic. For these reasons, *probabilistic encryption* was introduced in [Goldwasser and Micali 1982, 1984]. We shall later (in Chapter 5 and §2.2 of Chapter 6) see examples of probabilistic encryption.

On the negative side, probabilistic encryption systems sometimes are vulnerable to so-called *adaptive chosen-ciphertext attack* (see Exercise 11 of §3 of Chapter 5 and Exercise 6 of §2 of Chapter 6).

We shall next discuss two particularly important examples of public key cryptosystems – RSA and Diffie–Hellman/DSA. Both are connected with fundamental questions in number theory – factoring integers and discrete logarithms, respectively. Although the systems can be modified to perform most or all of the six tasks listed above, we shall describe protocols for only a few of these tasks (message transmission in the case of RSA, and key exchange and digital signature in the case of Diffie–Hellman).

§ 3. The RSA Cryptosystem

3.1 Encryption

Suppose that we have a large number of users of our system, each of whom might want to send a secret message to any one of the other users. We shall assume that the message units m have been identified with integers in the range $0 \leq m < N$. For example, a message might be a block of k letters in the Latin alphabet, regarded as an integer to the base 26 with the letters of the alphabet as digits; in that case $N = 26^k$. In practice, in the RSA system N is a number of between about 200 and 600 decimal digits.

Each user A (traditionally named Alice) selects two extremely large primes p and q whose product n is greater than N. Alice keeps the individual primes secret, but she publishes the value of n in a directory under her name. She also chooses at random an exponent e which must have no common factor with $p-1$ or $q-1$ (and probably has the same order of magnitude as n), and publishes that value along with n in the directory. Thus, her *public key* is the pair (n, e).

Suppose that another user B (Bob) wants to send Alice a message m. He looks up her public key in the directory, computes the least nonnegative residue of m^e modulo n, and sends Alice this value (let c denote this *ciphertext* value). Bob can perform the modular exponentiation $c \equiv m^e \pmod{n}$ very rapidly (see Example 3.5 of Chapter 2).

To decipher the message, Alice uses her secret *deciphering key* d, which is any integer with the property that $de \equiv 1 \pmod{p-1}$ and $de \equiv 1 \pmod{q-1}$. She can find such a d easily by applying the extended Euclidean algorithm to the two numbers e and l.c.m.$(p-1, q-1)$ (see Example 3.4 of Chapter 2; here "l.c.m." means "least common multiple"). One checks (see Exercise 1 below) that if Alice computes the least nonnegative residue of c^d modulo n, the result will be the original message m.

What would prevent an unauthorized person C (Catherine) from using the public key (n, e) to decipher the message? The problem for Catherine is that without knowing the factors p and q of n there is apparently no way to find a deciphering exponent d that inverts the operation $m \mapsto m^e \pmod{n}$. Nor does there seem to be any way of inverting the encryption other than through a deciphering exponent. Here I use the words "apparently" and "seem" because these assertions have not been proved. Thus, one can only say that *apparently* breaking the RSA cryptosystem is as hard as factoring n.

3.2 Hash Functions

Before discussing digital signatures, it is necessary to explain what a hash function is. Suppose that we are sending a message containing l symbols, and we would like our signature to be much shorter – say, about k symbols. Here is an informal definition of "hash".

Definition 3.1. A function $H(x)$ from the set of l symbols to the set of k symbols is called a *hash function* if $H(x)$ is easy to compute for any x, but
1) no one can feasibly find two different values of x that give the same $H(x)$ ("collision resistant"); and
2) given y in the image of H, no one can feasibly find an x such that $H(x) = y$ ("preimage resistant").

Much research has been devoted to both the theory and practical implementation of hash functions. We shall not dwell on this. In practice it is not very hard to find a function that satisfies the properties in Definition 3.1.

One of the main uses of a hash function is in digital signatures. Suppose that Bob sends Alice a long message x of l symbols. Both Alice and Bob are using the same hash function – and, in fact, there is no need for them to keep it secret from their adversary Catherine. After Bob sends Alice the message x, he appends the hash value $H(x)$. Alice would like to be certain that it was really Bob who sent the message x, and that Catherine did not alter his message before Alice received it. Suppose that she can somehow be certain that at least the appended $H(x)$ really did come from Bob. In that case all she has to do is apply the hash function to the message she received. If it agrees with $H(x)$, then she is happy: she knows that Catherine could not feasibly have tampered with x in such a way as to produce a distorted message x' such that $H(x') = H(x)$. The problem that remains is how Alice can be sure that $H(x)$ really came from Bob.

3.3 Signature

Here is how the last problem – how to be certain that $H(x)$ really came from Bob – can be solved using RSA. For convenience, choose k so that messages of length k are just small enough to make up one message unit; if the 26-letter Latin alphabet is being used, then k is the same as at the beginning of §3.1. After sending the message x, Bob computes the hash value $H = H(x)$. He does not simply send H to Alice, but rather first raises it to the power of his *deciphering exponent* d_{Bob} modulo n_{Bob}. Then Bob sends Alice the whole message x with $H' = H^{d_{Bob}} \pmod{n_{Bob}}$ appended, using Alice's enciphering exponent e_{Alice} and her modulus n_{Alice}. That is, he sends

$$\left(H^{d_{Bob}} \ (\text{mod } n_{Bob}) \right)^{e_{Alice}} \ (\text{mod } n_{Alice}) \ ,$$

where the notation $a \pmod{n}$ denotes the least nonnegative residue of a modulo n. After Alice deciphers the message, she takes the last message unit (which will look to her like gibberish rather than an intelligible plaintext message unit) and raises it to the power of Bob's *enciphering exponent* e_{Bob} modulo n_{Bob} in order to recover H. She then applies the hash function to the message, and verifies that the result coincides with H. Here the crucial observation is that Alice knows that *only Bob* would know the exponent that is inverted by raising to the e_{Bob}-th power modulo n_{Bob}. Thus, she knows that it really was Bob who sent her H. She also knows that it was he who sent the message x, which she received without any tampering.

It should be noted that this RSA signature has two other features besides simply allowing Alice to verify that it was in fact Bob who sent the message. In the first place, because the appended segment H' was encrypted along with the rest of the message, Bob's privacy is preserved; from the ciphertext an eavesdropper will not be able to find out who sent the message. In the second place, the signature ensures *non-repudiation*; that is, Bob cannot subsequently deny having sent the message.

§ 4. Diffie–Hellman and the Digital Signature Algorithm

The second landmark example of a public key cryptographic system is based on the discrete logarithm problem. First we define this problem.

Let $\mathbb{F}_p^* = (\mathbb{Z}/p\mathbb{Z})^* = \{1, 2, \ldots, p - 1\}$ denote the multiplicative group of integers modulo a prime p. (This group will be treated in more detail in §2 of Chapter 3.) Let $g \in \mathbb{F}_p^*$ be a fixed element (our "base"). The *discrete log problem* in $\mathbb{F}_p^*$ to the base g is the problem, given $y \in \mathbb{F}_p^*$, of determining an integer x such that $y = g^x$ (if such x exists; otherwise, one must receive an output to the effect that y is not in the group generated by g).

4.1 Key Exchange

The Diffie–Hellman key exchange works as follows. Suppose that Alice and Bob want to agree upon a large integer to serve as a key for some private key cryptosystem. This must be done using open communication channels – that is, any eavesdropper (Catherine) knows everything that Alice sends to Bob and everything that Bob sends to Alice. Alice and Bob first agree on a prime p and a base element g in $\mathbb{F}_p^*$. This has been agreed upon publicly, so that Catherine also has this information at her disposal. Next, Alice secretly chooses a random positive integer $k_{\text{Alice}} < p$ (of about the same magnitude as p), computes the least positive residue modulo p of $g^{k_{\text{Alice}}}$ (see Example 3.5 of Chapter 2), and sends this to Bob. Meanwhile, Bob does likewise: he sends $g^{k_{\text{Bob}}} \in \mathbb{F}_p^*$ to Alice, while keeping k_{Bob} secret. The agreed upon key will then be the integer

$$g^{k_{\text{Alice}} k_{\text{Bob}}} \in \mathbb{F}_p^* = \{1, 2, \ldots, p - 1\} \ ,$$

which Bob can compute by raising the integer he received from Alice to his secret k_{Bob}-power modulo p, and Alice can compute by raising the integer she received from Bob to the k_{Alice}-power modulo p. This works because in $\mathbb{F}_p^*$ we have

$$g^{k_{\text{Alice}} k_{\text{Bob}}} = \left(g^{k_{\text{Alice}}}\right)^{k_{\text{Bob}}} = \left(g^{k_{\text{Bob}}}\right)^{k_{\text{Alice}}} \ .$$

The problem facing the adversary Catherine is the so-called *Diffie–Hellman problem*: Given g, g^{k_A}, $g^{k_B} \in \mathbb{F}_p^*$, find $g^{k_A k_B}$. It is easy to see that anyone who can solve the discrete log problem in $\mathbb{F}_p^*$ can then immediately solve the Diffie–Hellman problem as well. The converse is not known. That is, it is conceivable (though thought to be unlikely) that someone could invent a way to solve the Diffie–Hellman problem without being able to find discrete logarithms. In other words, breaking the Diffie–Hellman key exchange has not been *proved* to be equivalent to solving the discrete log problem (although some recent partial results in this direction support the conjectured equivalence of the two problems; see [Boneh and Lipton 1996]). For practical purposes it is probably safe to assume that the Diffie–Hellman key exchange is secure provided that the discrete logarithm problem is intractable.

4.2 The Digital Signature Algorithm (DSA)

In 1991 the U.S. government's National Institute of Standards and Technology (NIST) proposed a Digital Signature Standard (DSS) based on a certain Digital Signature Algorithm (DSA). The role of DSS is expected to be analogous to that of the older Data Encryption Standard (DES): it is supposed to provide a standard digital signature method for use by government and commercial organizations. But while DES is a classical ("private key") cryptosystem, in order to construct digital signatures it is necessary to use public key cryptography. NIST chose to base their signature scheme on the discrete log problem in a prime finite field $\mathbb{F}_p$. The DSA is very similar to a signature scheme that was originally proposed in [Schnorr 1990]. It is also similar to a signature scheme in [ElGamal 1985a]. We now describe how the DSA works.

To set up the scheme (in order later to be able to sign messages), each user Alice proceeds as follows:

1) she chooses a prime q of about 160 bits (to do this, she uses a random number generator and a primality test);
2) she then chooses a second prime p that is $\equiv 1 \pmod{q}$ and has about 500 bits (more precisely, the recommended number of bits is a multiple of 64 between 512 and 1024);
3) she chooses a generator g of the unique cyclic subgroup of $\mathbb{F}_p^*$ of order q (she does this by computing $g_0^{(p-1)/q} \pmod{p}$ for a random integer g_0; if this number is not equal to 1, it will be a generator);
4) she takes a random integer x in the range $0 < x < q$ as her secret key, and sets her public key equal to $y = g^x \pmod{p}$.

Now suppose that Alice wants to sign a message. She first applies a hash function to her plaintext (see §3.2), obtaining an integer H in the range $0 < H < q$. She next picks a random integer k in the same range, computes $g^k \pmod{p}$, and sets r equal to the least nonnegative residue modulo q of the latter number (that is, g^k is first computed modulo p, and the result is then reduced modulo the smaller prime q). Finally, Alice finds an integer s such that $sk \equiv H + xr \pmod{q}$. Her signature is then the pair (r, s) of integers modulo q.

To verify the signature, the recipient Bob computes $u_1 = s^{-1}H \pmod{q}$ and $u_2 = s^{-1}r \pmod{q}$. He then computes $g^{u_1}y^{u_2} \pmod{p}$. If the result agrees modulo q with r, he is satisfied. (See Exercise 2 at the end of the chapter.)

This signature scheme has the advantage that signatures are fairly short, consisting of two numbers of 160 bits (the magnitude of q). On the other hand, the security of the system seems to depend upon intractability of the discrete log problem in the multiplicative group of the rather large field $\mathbb{F}_p$. Although to break the system it would suffice to find discrete logs in the smaller subgroup generated by g, in practice this seems to be no easier than finding arbitrary discrete logarithms in $\mathbb{F}_p^*$. Thus, the DSA seems to have attained a fairly high level of security without sacrificing small signature storage and implementation time.

There is a variant of DSA using elliptic curves that might be even harder to break than the finite-field DSA described above. This elliptic curve version will be discussed in Chapter 6.

§ 5. Secret Sharing, Coin Flipping, and Time Spent on Homework

5.1 Secret Sharing

Suppose that you want to give enough information to a group of people so that a secret password – which we think of as an integer N – can be determined by any group of k of them; but if only $k - 1$ collaborate, they won't get anywhere. Here is a way to do this. Choose an arbitrary point $P = (x_1, \ldots, x_k)$ in the Euclidean space $\mathbb{R}^k$, where the x_i are integers and $x_1 = N$. Give each person in the group a single linear equation in k variables that is satisfied by P. Each equation determines a hyperplane in $\mathbb{R}^k$ that contains P. Choose your equations so that any k of them are linearly independent. (In other words, the coefficient matrix of any k of the equations has nonzero determinant.) Then any k people can solve the corresponding $k \times k$ system of linear equations for the point P. But $k - 1$ equations determine a line, and so give no information about the first coordinate of P. (Here we're assuming that the line is not contained in the first coordinate hyperplane; a judicious choice of the linear equations will guarantee this.)

Another method of secret sharing is to choose a prime p for each person, and give him or her the value of the least nonnegative residue of N modulo p. N must be in a range where it can be uniquely recovered (using the Chinese Remainder Theorem, see Exercise 9 in §3 of Chapter 2) from its set of remainders modulo p for k values of p, but not from its remainders for $k - 1$ values of p.

5.2 Bit Commitment

Suppose that Alice and Bob want to decide who gets a certain advantage – for example, who gets to play white in a chess match, or whose city gets to be the home team for the volleyball championship game. They can determine this by flipping a coin, provided that they are in the same physical location and both trust the fairness of the coin. Alternatively, they can "shoot fingers" – again, supposing that they are in the same place. That is, one of them (say, Alice) calls out "evens". Then they simultaneously throw out either one or two fingers. If the sum of the fingers is even (in other words, 2 or 4), then Alice wins. If the sum of the fingers is odd (in other words, 3), then Bob wins.

A cryptographic problem arises when Alice and Bob are far away from one another, and when they must act sequentially rather than at the same instant. In that case they need a procedure for *bit commitment*.

Definition 5.1. A *bit commitment* protocol is a procedure whereby Alice puts a secret bit (that is, either 0 or 1) in an "envelope", to be revealed after Bob guesses which bit it is. Bob must not be able to increase his odds of guessing the right bit beyond 50%, and Alice must not be able to change the bit after she puts it in the "envelope".

Here is an example of a bit commitment protocol. Suppose that Alice and Bob each have a "machine" that takes in a string of m bits and outputs a string of n bits. The machine should be constructed so as to be rather complicated, for all practical purposes operating much like a random function from $\{0, 1\}^m$ to $\{0, 1\}^n$. For instance, the machine might be a large Boolean circuit made up of and-gates, or-gates, and not-gates. After constructing their circuits, Alice and Bob each send the other a copy of his or her circuit. Next, Alice secretly chooses a random sequence of m bits. She puts the sequence through both her and Bob's circuits, and adds the resulting vectors modulo 2 (this is called the XOR operation, denoted $\oplus$: $0 \oplus 0 = 1 \oplus 1 = 0$ and $0 \oplus 1 = 1 \oplus 0 = 1$). She sends the sum to Bob. Bob now tries to guess the *parity* of her input, that is, whether there were an odd or even number of 1's in it. If he guesses incorrectly, Alice must prove to him that he is wrong by revealing her input – at which point Bob can verify that the XOR of the outputs of the two circuits is in fact what Alice sent him before. That is, the message that Alice sent him prevents her from changing her input after Bob guesses its parity.

Note that one needs certain conditions in order for this Boolean circuit protocol to be a fair bit commitment scheme. The circuits must be complicated enough so that (1) Bob cannot somehow invert them and recover the input, and (2) Alice cannot find two different inputs of opposite parity that lead to the same output. (Compare with the two properties in Definition 3.1.)

5.3 Concealing Information

Suppose that a teacher wants to find out the average number of hours per week that the students are spending on homework. If each student were asked to reveal this number, there would be many distorted answers, for at least two reasons. First, those who devote hardly any time to their homework might not want the teacher to know this. Second, those who spend a lot of time on their homework might not want the other children to know, for fear of seeming odd – a "nerd" or "teacher's pet".* Note that the teacher is interested in knowing only the average, not any of the individual values.

Here is a procedure for determining the average while concealing all individual values. Starting with Alice, the children form a chain going around the classroom and finally returning to Alice. Alice secretly chooses a number at random, adds to

---

* The second reason, which is based on the psychology of American children, might not apply in countries where children do not grow up surrounded by an anti-intellectual popular culture.

it her figure for the number of hours she spends on homework, and whispers the sum to the second student (Beatrice). Beatrice adds her number of hours to the number she received from Alice, and whispers the sum to Catherine. Catherine adds the number of hours she spends on homework and passes the sum to the next child, and so on. Finally, the sum is passed back to Alice, who subtracts her secret number and reveals the result. The teacher divides this total by the number of students to find the average. No one has learned anyone else's individual value, but everyone knows the average.

§ 6. Passwords, Signatures, and Ciphers

Public key cryptosystems for passwords, for signatures, and for encryption all use one-way functions, but in somewhat different ways. Roughly speaking, any one-way function can be used for passwords, whereas encryption requires the presence of a "trapdoor". Signatures are somewhere in between. We now explain this.

Recall how a password system works (see the end of §1). Let $x \mapsto y = f(x)$ be a function that is easy to compute but computationally impossible to invert – that is, in practice it is not feasible to compute the inverse function $g = f^{-1}$. Users' passwords are values of x in the domain of the function $f(x)$. To keep the list of passwords out of the hands of intruders (hackers), the computer does not store these passwords x. Rather, under each user's name it stores the value $f(x)$ that is obtained by applying the function f to her password x. Any time she wants to log in, she types her password x. The computer calculates $f(x)$, matches it with the $f(x)$ under her name, grants her access to the system, and then deletes any record of x.

Encryption also uses a one-way function f. This function goes from plaintext message units x to ciphertext message units y, and it depends on the addressee's encryption key. However, not any one-way function f will work. One needs to use an f that is a one-way function from the perspective of the general public, but is a two-way function (that is, both f and its inverse $g = f^{-1}$ are easy to compute) from the perspective of the addressee, who has an additional piece of information, namely, the decryption key. In the case of RSA, for example, the additional information can be either a decryption exponent or the factorization of the modulus n (from which a decryption exponent can easily be found). That is, a *trapdoor one-way function* is a function whose one-way status depends on keeping some piece of information secret. There are many one-way functions – for example, Purdy's polynomial from $\mathbb{F}_p$ to $\mathbb{F}_p$ at the end of §1 – that are not trapdoor one-way functions, because even the creators of the system have no advantage over anyone else in inverting the function. That is, there's no additional amount of information that anyone knows that could give a method for finding $x = f^{-1}(y)$.

For a signature system one needs something more than for a password system, but not a full trapdoor in the sense of the last paragraph. We want a procedure for Alice to verify that the message m that she received, supposedly from Bob, really did come from Bob. Bob wants to convince her that only he could have sent her the message. Let $H(m)$ be the "hashed message". This is a much shorter sequence of symbols. The function H must have the property that it is computationally impossible in practice to find two different messages m and m' such that $H(m) = H(m')$. In addition, given a y in the image of H, it must not be feasible to find a message m such that $H(m) = y$. The hash function H is publicly known – anyone can compute $H(m)$ for any message m.

Let $y = f(x)$ be a function that is defined implicitly, in the sense that for any given x and y it is easy to verify whether or not $y = f(x)$. (This notion is familiar from calculus – for example, the equation $e^{xy} = y - x$ defines a curve passing through the point $(0, 1)$, and near $x = 0$ it gives a single-valued function of x, but this function $y = f(x)$ cannot be expressed in closed form.) Suppose that Alice knows that only Bob has an additional piece of information needed to compute the inverse function $x = g(y)$. Then if Bob sends Alice the value $H' = g(H(m))$, she can verify that $H(m) = f(H')$, even though she might not have been able to compute $f(H')$ and certainly could not have computed $g(H(m))$. That is, all Alice has to do to become convinced that Bob sent the message m (and that m was not tampered with before she received it) is to verify that $H(m) = f(H')$.

To summarize, for a password system we need a function that is easy in one direction and impossible in the other direction. For an encryption system we need a function that is easy in one direction and impossible in the inverse direction unless we know an additional secret piece of information, in which case it is easy in both directions. For a signature system our function f is impossible in the inverse direction unless we know an additional secret piece of information (in which case it is easy in that direction), and it must be easy to verify whether or not $y = f(x)$ for any given x and y.

In the case of RSA, the one-way function that we used for signatures was the same as the one-way function that we used for encryption. However, in some situations it might be advantageous to use a one-way function for signatures that does not satisfy the more stringent requirements for encryption. We shall give an example in §3 of Chapter 4.

§ 7. Practical Cryptosystems and Useful Impractical Ones

7.1 Criteria for a Cryptosystem to be Practical

The most obvious quality one looks for in a cryptosystem is *security*. That is, it must not be feasible for an adversary to break the system. In §5 of Chapter 2 we give a more precise definition of what it means to break (or "crack") a cryptosystem. The science (or art) of trying to break cryptosystems is called *cryptanalysis*.

One can never be sure – in the sense of a rigorous mathematical proof – that a public key cryptosystem cannot feasibly be broken. The best one can hope for is to have a large amount of empirical evidence that

1) the system cannot be cracked without solving a certain mathematical problem, and

2) there is no method that anyone knows for solving this mathematical problem in a reasonable length of time, provided that certain conditions are met.

For example, in the case of RSA (1) it is widely believed that there is no way to break the system without factoring the modulus n; and (2) none of the state-of-the-art factoring algorithms and computer facilities can factor a suitably chosen n in a reasonable length of time if n has at least 200 digits.

But one has to be cautious. Sometimes attacks are found that might compromise the cryptosystem without solving the mathematical problem directly. For example, it turns out that one can sometimes get valuable information by simply timing how long Alice's computer takes to perform the steps in RSA or some other system (see [Kocher 1996]). Moreover, some implementations of supposedly secure cryptosystems have been broken because the designers had "cut corners".

In addition, we have to be sure that condition 2) above holds not simply because few people have attempted to solve the problem. A cryptosystem should be based on a problem that has been widely studied both theoretically and computationally. One of the main reasons for the popularity of RSA and the confidence that people have in it is that its security is based on a famous problem that has interested mathematicians for centuries and has been seriously studied for decades – integer factorization.

A second basic practicality issue is *efficiency*. For instance, one might want to send vast amounts of encrypted data in just a few seconds. In general, public key systems for message encryption are much slower than private key systems. They are fast enough when the message is not extremely long. Even in cases when the volume of data is great and one needs a private key system, public key cryptography is extremely useful in exchanging and managing the keys for such a system.

Besides speed of operation, one might also be interested in economy of space. For instance, so-called *smart cards* have very limited memory. This means that it is desirable to have public key cryptosystems that (1) use fairly simple algorithms that can be built into a small chip and (2) only need keys of relatively small bit-length. It is for this reason that elliptic curve cryptosystems have been proposed for such purposes (see Chapter 6). In the case of digital signatures, some of the hidden monomial cryptosystems (see Chapter 4) might have a similar advantage. If all known algorithms for breaking a given cryptosystem require fully *exponential time* (see Chapter 2) – this is the case for the elliptic curve and the hidden monomial systems – then one is likely to be able to use short keys while maintaining a high level of security.

In addition, it is important to have a reasonably efficient algorithm for generating keys. In cases when virtually any random integer in a certain range will suffice, this is relatively easy. However, if the integers (or other mathematical objects) needed for the keys must satisfy some additional properties, then we must put some thought into creating efficient and reliable algorithms that generate possible keys and test them for suitability.

A final remark should be made on the question of security. Security is a relative notion. A cryptosystem that is secure for certain purposes (e.g., keeping a message private for a few hours until it can become public knowledge) might not be secure if the demands are more stringent (keeping information confidential for the next 25 years). A cryptosystem that cannot be broken using our present-day knowledge and technology might succumb to cryptanalysts in the 23rd century or to advanced extraterrestrial mathematicians from a distant star system.

In the other direction, a cryptosystem that would never win the respect of a professional cryptographer might be quite secure in the face of adversaries who do not have advanced mathematical training. Such a system could be used, for example, in children's games and high school math clubs. An example is given in the exercises below.

7.2 The "Unreasonable Effectiveness" of Theory

In a famous article Eugene Wigner wrote that

> ... the enormous usefulness of mathematics in the natural sciences is something bordering on the mysterious and...there is no rational explanation for it. [Wigner 1960]

Wigner was speaking primarily of physics, but a similar observation could be made about some other branches of science, in particular cryptography.

For example, the study of complexity classes (see Chapter 2) is part of theoretical computer science – that is, essentially it is a branch of pure mathematics. It is almost miraculous – in the sense that Wigner uses the term – that it has much to say about practical matters. For instance, the notion of a "polynomial time" problem is fundamental in complexity theory (see Definition 4.1 of Chapter 2). However, there is no *a priori* reason why a polynomial time problem should be easier to solve in practice than a problem that is not polynomial time. For instance, suppose that the best algorithm we know for the problem $\mathcal{P}_1$ has running time of order Cn^{50}, where n is the *length* of the input (the number of symbols in the input) and C is a constant; and suppose that the best algorithm we know for another problem $\mathcal{P}_2$ has running time of order $Cn^{\ln n}$. Then $\mathcal{P}_1$ is a polynomial time problem, whereas $\mathcal{P}_2$ might not be. Nevertheless, we can solve $\mathcal{P}_2$ *much* faster than $\mathcal{P}_1$, unless the input involves more than a billion trillion symbols. Of course, if $n \approx 10^{21}$, then both algorithms will be completely impractical. Thus, for all practical purposes $\mathcal{P}_2$ is much easier to solve than the polynomial time problem $\mathcal{P}_1$.

It is curious that the situation just described almost never occurs. Almost all of the polynomial time problems one encounters in practice have algorithms whose running times are very reasonable – they are bounded by Cn^k with k small (usually 2, 3 or 4) and C not too large. The theoretical class of polynomial time problems seems to include a large proportion of the problems of practical interest that can be quickly solved on today's computers, and seems to include very few problems that cannot be. (See §4.2 of Chapter 2 for more discussion of this point.)

A second example of the "unreasonable effectiveness" of theory is that ancient topics in number theory – the distribution of prime numbers and the factoring of composite numbers – are at the heart of the most popular practical public key cryptosystem, RSA. Elliptic curve cryptography provides a third example of the unexpected connections between practical questions and areas of basic research that were once thought to be "math for math's sake" (see especially §3 of Chapter 6). Elliptic curves have a rich history in number theory, particularly in the branch known as *arithmetic algebraic geometry*. This history reached its culmination when Andrew Wiles stunned the mathematical world with a proof of Fermat's Last Theorem based on an elaborate study of the properties of these curves. Wiles' work is a purely intellectual achievement, without practical consequences as far as anyone knows. It is remarkable that the same mathematical objects that Wiles worked with can also be used to construct what many think are among the most efficient and secure public key cryptosystems.

7.3 The Need for Impractical Cryptography

From a narrow point of view an idea for a cryptosystem is worthless unless the necessary conditions discussed in §7.1 are satisfied. That is, one must have algorithms to set up the system (generate keys) that all but guarantee unbreakability and algorithms to implement the cryptographic procedures that are at least as efficient as those of competing systems.

Moreover, one who adheres to this restrictive viewpoint can argue that there is no real need for a large number of cryptosystems. In fact, in the real world it is preferable to reach a consensus favoring a small selection of the best available systems. In practice, the way this works is that the leading professional organizations adopt a formal set of "standards". Such standards are a necessity if one wants a high level of quality control and interoperability. One can argue that cryptographic research is worthwhile only insofar as it will ultimately lead to an improved set of standards or additional standards for newly developed cryptographic purposes.

But one can also look at cryptography from a broader perspective. The subject is closely connected with other areas of science, such as (1) computational mathematics, (2) complexity theory, and (3) the theory of games. A cryptographic idea that may never lead to a new standard in practical cryptography might nonetheless be worth thinking about because:

1) it might give rise to some interesting questions in theoretical mathematics, and give mathematicians a new slant on old theories;
2) it might shed light on the interrelationships between complexity classes, and suggest new directions of research in theoretical computer science;
3) it might provide a natural and entertaining way to popularize mathematics and computer science among the general public;
4) it might lead to an effective teaching tool to use with children.

In connection with 3) and 4), we make the following definition.

Definition 7.1. *Kid Krypto* is the development of cryptographic ideas that are accessible and appealing (and moderately secure) to those who do not have university-level mathematical training.

See Exercise 4 below for "kid-RSA"; for more examples and discussion see [Fellows and Koblitz 1994a] and [Koblitz 1997].

Exercises for Chapter 1

1. Suppose that p and q are distinct primes, and d and e are two positive integers such that $ed \equiv 1 \pmod{\text{l.c.m.}(p-1, q-1)}$. Let $n = pq$. Prove that for any integer m one has $m^{ed} \equiv m \pmod{n}$.

2. In the DSA, explain why (a) Bob expects $g^{u_1} y^{u_2}$ to agree modulo q with r, and (b) if they agree, he should be satisfied that it really was Alice who sent the message.

3. Explain in more detail how to share a secret using the Chinese Remainder Theorem. (See Exercise 9 in §3 of Chapter 2.)

4. Suppose that the following cryptosystem is introduced among secondary school students who have learned how to reduce numbers modulo a positive integer n and how to convert numbers from one base to another (in particular, how to work with blocks of letters regarded as integers to the base 26). To set up the system, each student (Alice) chooses any two integers a and b, sets $M = ab - 1$, then chooses two more integers a' and b', and finally sets

$$e = a'M + a, \quad d = b'M + b, \qquad n = \frac{ed - 1}{M} = a'b'M + ab' + a'b + 1 .$$

Her public key is (n, e), and her private key is d. To send Alice a plaintext m, one uses the map $c \equiv em \pmod{n}$; Alice deciphers the ciphertext by multiplying by d modulo n.

   (a) Verify that the decryption operation recovers the plaintext.
   (b) Show how to make digital signatures.
   (c) Show how the Euclidean algorithm (see §3.3 of Chapter 2) completely breaks the system.
   (d) Can you prove that the ability to crack this cryptosystem (for any choice of a, b, a', b') implies the ability to solve the equation $xr + ys = 1$ for any two relatively prime integers r and s? Could there be a way to crack the system without essentially rediscovering a version of the Euclidean algorithm?
   (e) Suppose that you are teaching an introductory number theory course. Instead of presenting the Euclidean algorithm to students on a silver platter, you give them the above cryptosystem, in the hope that it will give them an incentive to discover the Euclidean algorithm on their own, and thereby better appreciate its power and beauty. Would this work as a pedagogical method? (See [Koblitz 1997].)

Chapter 2. Complexity of Computations

§1. The Big-O Notation

Suppose that $f(n)$ and $g(n)$ are functions of the positive integers n which take *positive* (but not necessarily integer) values for all n. We say that $f(n) = O(g(n))$ (or simply $f = O(g)$) if there exists a constant C such that $f(n)$ is always less than $C \cdot g(n)$. For example, $2n^2 + 3n - 3 = O(n^2)$ (namely, it is not hard to prove that the left side is always less than $3n^2$, so 3 can be chosen as the constant C in the definition).

In practice, when we use the big-O notation we do not care about what the functions f and g are like for small values of n. For this reason, we shall actually make a somewhat broader definition of the notation.

Definition 1.1. Suppose that for all $n \geq n_0$ the two functions $f(n)$ and $g(n)$ are defined, take positive values, and for some constant C satisfy the inequality $f(n) \leq C \cdot g(n)$. Then we say that $f = O(g)$.

Remarks. 1. Despite the equality sign in the notation $f = O(g)$, we should think of big-O as conveying "less than" type information. For example, it is correct to write $n\sqrt{n} = O(n^2)$, but it is incorrect to write $n^2 = O(n\sqrt{n})$.

2. Of course, the variable is not always called n. In any given situation we must understand clearly what letter is standing for the variable – there might be several letters in use which are standing for constants. Example 1.3 below illustrates the importance of knowing what letter is the variable.

3. In practice, we will use this notation only when $g(n)$ is a simpler function than $f(n)$ and does not increase a whole lot faster than $f(n)$ – in other words, when $g(n)$ provides a "good idea" (a "pretty close upper bound") for how fast $f(n)$ is increasing. The following statements, all of which are mathematically correct, are *not* useful in practice: (1) $n^2 = O(n^3 + n^2 \ln n + 6683)$; (2) $n^2 = O(e^{(n^2)})$; (3) $e^{-n} = O(n^2)$.

4. Suppose that $f(n)$ is a sum of terms, one of which is much larger than the others when n is large. If we let $g(n)$ denote that "dominant term", then we can write $f(n) = O(g(n))$. For example, if $f(n)$ is *any* polynomial of degree 3 (with positive leading coefficient), then $f(n) = O(n^3)$. Similarly, if $f(n)$ is a polynomial of degree d (where d is any constant, and the coefficient a_d of n^d is positive), then $f(n) = O(n^d)$. The leading term $a_d n^d$ is the "dominant term".

5. If we are given $f(n)$ and make a good choice of $g(n)$ – that is, we choose $g(n)$ to be a simpler function such that $f = O(g)$ but $g(n)$ does not increase much faster than $f(n)$ – then the function $g(n)$ is useful in giving us an idea' of how a big increase in n will affect $f(n)$. For example, we can interpret the statement $f(n) = O(n)$ to mean "if n doubles in size, then $f(n)$ will also roughly double in size". (Notice that the value of the constant C in the definition of the big-O notation does not affect the truth of this statement. For example, if $f(n)$ is equal to roughly $2n$, then the words in quotes are true; and they are also true if $f(n)$ is equal to roughly $200n$.) We can interpret the statement $f(n) = O(n^2)$ to mean "if n doubles, then $f(n)$ will increase roughly by a factor of 4". The statement $f(n) = O(n^3)$ would mean that $f(n)$ increases roughly by a factor of 8.

We can interpret the statement $f(n) = O(2^n)$ to mean "if n increases by 1, then $f(n)$ will approximately double in size". For example, the statement $5n^3 + \binom{n}{4} + 2^n = O(2^n)$ means that for large n the expression on the left roughly doubles if n is increased by 1, since its dominant term is 2^n.

6. If $f(n)$ and $g(n)$ are two positive functions for $n \geq n_0$, and if

$$\lim_{n \to \infty} \frac{f(n)}{g(n)} = \text{any constant} ,$$

then it is not hard to show that $f = O(g)$. If the limit is zero, then it is still correct to write $f = O(g)$; but in that case we also say that "f is little-o of g" and we write $f = o(g)$. This means that $f(n)$ is much smaller than $g(n)$ when n is large.

7. Sometimes we want to make a more precise statement about the relationship between $f(n)$ and $g(n)$. For example, we might say that as n increases the percent error in using $g(n)$ in place of $f(n)$ goes to zero. That is, we might have

$$\lim_{n \to \infty} \frac{f(n)}{g(n)} = 1 .$$

In that case we write

$$f \asymp g$$

and say that "$f(n)$ is asymptotically equal to $g(n)$ for large n". For example, if $f(n)$ is any polynomial with leading term $a_d n^d$, then $f(n) \asymp a_d n^d$. Another example is the

Prime Number Theorem.

$$\pi(n) \asymp \frac{n}{\ln n} ,$$

where $\pi(n)$ denotes the number of prime numbers less than or equal to n.

8. There are two other commonly used symbols that are closely related to big-O: Ω and Θ. The notation $f = \Omega(g)$ means exactly the same thing as $g = O(f)$. The notation $f = \Theta(g)$ means that both $f = O(g)$ and $f = \Omega(g)$; in other words, there exist positive constants C_1, C_2, and n_0 such that $C_1 g(n) \leq f(n) \leq C_2 g(n)$ for $n \geq n_0$.

9. These symbols are often used in the middle of formulas rather than right after an equal sign. For example, if we say that a function is $n^{O(\ln \ln n)}$, we mean that there exists a constant C such that for $n \geq n_0$ the function is $\leq n^{C \ln \ln n}$. If we say that a function is $n^{\Theta(1)}$, we mean that for $n \geq n_0$ it is wedged between two constant powers of n.

Example 1.1. If ε stands for any positive constant at all, no matter how small (e.g., $\varepsilon = 0.001$), then $\ln n = O(n^\varepsilon)$. (We actually can write $\ln n = o(n^\varepsilon)$ as well.) To convince yourself of this, try putting in very, very large numbers for n – for example, extremely large powers of 2 – in the two functions $\ln n$ and $n^{0.001}$. For instance, setting $n = 2^{1000000}$ gives $1000000 \ln 2 = 693147.18 \cdots$ for $\ln n$ on the left and 2^{1000} (which is a 302-digit number) for n^ε on the right. More precisely, using l'Hôpital's rule it is easy to show that

$$\lim_{n \longrightarrow \infty} \frac{\ln n}{n^\varepsilon} = 0 \ .$$

Example 1.2. Let $f(n)$ be the number of base-b digits in n, that is, the length of the number n when it is written to the base b. Here b is constant, and n is our variable. Then

$$f(n) = 1 + \left[\log_b n\right] = 1 + \left\lfloor \frac{\ln n}{\ln b} \right\rfloor \ ,$$

where [ ] denotes the greatest integer function. Since b (and hence $\ln b$) is a constant, we conclude that $f(n) = O(\ln n)$. Roughly speaking, "the number-of-digits function behaves like a logarithm".

Example 1.3 (see Remark 2 above). This example shows the importance of being clear about what the variable is when one uses big-O, little-o, and asymptotic equality. We consider the sum of the first n positive integers raised to the k-th power: $\sum_{i=1}^{n} i^k$. If we are considering k to be constant and letting n get large, then we have

$$f(n) = \sum_{i=1}^{n} i^k \asymp \frac{n^{k+1}}{k+1} \ ,$$

that is, $f \asymp g$ with $g(n) = n^{k+1}/(k+1)$. (To see this, show that $\frac{1}{n^{k+1}} f(n)$ is equal to the n-th Riemann sum for the integral $\int_0^1 x^k dx$, and conclude that $\lim_{n \longrightarrow \infty} f(n)/g(n) = 1$.) On the other hand, if we regard n as constant and k as the variable that gets large, then the statement

$$f(k) = \sum_{i=1}^{n} i^k \asymp \frac{n^{k+1}}{k+1}$$

is false. For example, if n is the constant 2, then this statement says that $1 + 2^k \asymp \frac{2}{k+1} 2^k$, which is not true. Even the weaker statement $1 + 2^k = O\left(\frac{2}{k+1} 2^k\right)$ is false.

Final Remark on Big-O. Often we consider functions of more than one variable, say, $f(m, n)$. In that case the notation $f = O(g)$ is used when $g(m, n)$ is a simple expression involving m and n such that there exists a constant C with $f(m, n) \leq C \cdot g(m, n)$ provided that $m \geq m_0$ and $n \geq n_0$ (in other words, we are not interested in small values of the variables).

Example 1.4. Let $f(m, n)$ be the number of points with integer x- and y-coordinates that are contained inside an ellipse in the xy-plane with semimajor axis m and semiminor axis n. Then $f(m, n) = O(mn)$. In fact, $f(m, n)$ is approximately equal to the area of the ellipse, which is πmn, but the exact value depends on how the ellipse is situated in the plane. In any case it is not hard to show that $f(m, n) \leq 4mn$ if m and n are large, and thus $f(m, n) = O(mn)$.

Exercises for § 1

For each of the $f(n)$ in Exercises 1–11, give the letter of the best estimate among the following:

(a) $f(n) = O(\ln n)$;   (b) $f(n) = O(\ln^2 n)$;   (c) $f(n) = O(\ln^3 n)$;

(d) $f(n) = O(n)$;   (e) $f(n) = O(n^2)$;   (f) $f(n) = O(n^3)$;

(g) $f(n) = O(2^n)$;   (h) $f(n) = O(n!)$;   (i) $f(n) = O(n^n)$.

1. $\binom{n}{3}$.
2. $10 \ln^3 n + 20n^2$.
3. The number of monomials in x, y, z of total degree at most n.
4. The number of polynomials in x of degree at most n whose coefficients are 0 or 1.
5. The number of polynomials in x of degree at most $n - 1$ whose coefficients are integers between 0 and n.
6. The area of a fixed shape after it's magnified by a factor of n.
7. The amount of memory space a computer requires to store the number n.
8. The amount of memory space a computer requires to store n^2.
9. The sum of the first n positive integers.
10. The sum of the squares of the first n positive integers.
11. The number of bits (base-2 digits) in the sum of the squares of the first n positive integers.

For each of the $f(m, n)$ given below, find the best simple function $g(m, n)$ such that $f = O(g)$.
12. $(m^2 + 2m - 3)(n + \ln^2 n + 14)$.
13. $2m \ln^2 n + 3m^2 \ln n$.
14. The largest n-digit number to the base m.
15. The maximum number of circles of radius $1/n$ that fit into a circle of radius m without overlapping.

§ 2. Length of Numbers

From now on, unless otherwise stated, we shall assume that all of our numbers are written in binary, and all arithmetic is performed to the base 2. Throughout this book we shall use the notation log to mean $\log_2$ and ln to mean $\log_e$.

By the "length" of an integer we mean the number of bits (binary digits) it has. Recall that

$$\text{length}(n) = 1 + \left\lfloor \log_2 n \right\rfloor = 1 + \left\lfloor \frac{\ln n}{\ln 2} \right\rfloor ,$$

and this can be estimated by $O(\ln n)$.

In this section we shall discuss how to estimate the length of a number which is arrived at by various arithmetic processes, starting with "input" of known length (or whose length is assumed to be no greater than a known bound).

Example 2.1. What is the length of the number that is obtained by
(a) adding together, and
(b) multiplying together
n positive integers each of which has length at most k?

Solution. To answer this question we have to think about how adding and multiplying affect the length of numbers. It is easy to see that *the sum of two numbers has length either equal to the length of the larger number or else equal to 1 plus the length of the larger number.*

If we add n numbers each of length at most k – that is, each less than 2^k – then the sum will be less than $n2^k$. Hence, the length of the sum will be at most $k + \text{length}(n)$.

To deal with multiplication, we use the fact that a number m of length k satisfies: $2^{k-1} \le m < 2^k$. Thus, if m_1 has length k and m_2 has length l, we can multiply the two inequalities

$$2^{k-1} \le m_1 < 2^k$$
$$2^{l-1} \le m_2 < 2^l$$

to get: $2^{k+l-2} \le m_1 m_2 < 2^{k+l}$, from which it follows that $\text{length}(m_1 m_2)$ *is equal either to the sum of the lengths of m_1 and m_2 or else to 1 less than the sum of the lengths of m_1 and m_2.* Roughly speaking, when we multiply numbers, their lengths add together. In other words, the lengths of numbers behave like logarithms. (See Example 1.2.)

Now suppose that we want to multiply together n k-bit numbers $m_1, \ldots, m_n$. (For example, the m numbers might all be the same, in which case we're raising a k-bit number to the n-th power.) If we multiply together all n inequalities

$$2^{k-1} \le m_i < 2^k , \qquad i = 1, \ldots, n ,$$

then we find that

$$2^{nk-n} \le \prod m_i < 2^{nk} ,$$

so that the length of the product is between $nk - (n-1)$ and nk.

Usually we're not interested in the exact length, but only in a bound for the length. In that case we can say simply that *multiplying together n numbers of length at most k results in a number of length at most nk.*

A similar discussion applies to subtraction and division (see Exercise 1 below).

Example 2.2. Find the length of $n!$.

Solution. Here what we want is a simple estimate for the length of $n!$ in the form $O(g(n))$. Notice that none of the n numbers that are multiplied together in $n!$ has length longer than length(n). So we can apply the statement in italics above to conclude that: length$(n!) \leq n(\text{length}(n)) = O(n \ln n)$.

One might object that $O(n \ln n)$ is not the best possible estimate, since, after all, most of the numbers multiplied together in $n!$ are quite a bit less than n. However, notice that most of the numbers from 1 to n have length not a whole lot less than the length of n. In Exercise 4 below we shall see that length$(n!)$ not only is less than $Cn \ln n$, but also is *greater* than some other constant C' times $n \ln n$. That is, length$(n!) = \Theta(n \ln n)$.

Exercises for §2

1. Suppose that a k-bit integer a is divided by an l-bit integer b (where $l \leq k$) to get a quotient q and a remainder r:

$$a = qb + r , \qquad 0 \leq r < b .$$

What is the length of q?

2. In each case estimate the length of the number indicated. Express your answer using the big-O notation with a simple function $g(n)$, $g(k)$, $g(n, k)$, etc. Here g must be expressed using the letters given in the statement of the problem.
(a) The sum of n numbers, each of length at most k.
(b) $n^4 + 25n^2 + 40$.
(c) A polynomial in n of degree k: $a_k n^k + a_{k-1} n^{k-1} + \cdots + a_1 n + a_0$, where k and the a_i are integer constants.
(d) The product of all prime numbers of k or fewer bits.
(e) $(n^2)!$.
(f) The n-th Fibonacci number. (The Fibonacci numbers are defined by setting $f_1 = 1$, $f_2 = 1$, and $f_{n+1} = f_n + f_{n-1}$ for $n = 2, 3, \ldots$.)

3. Find a simple function $g(n)$ such that the length of the n-th Fibonacci number is asymptotically equal to $g(n)$.

4. Show that at least $n/2$ of the numbers $1, 2, 3, \ldots, n$ have length equal to or greater than $\log_2 n - 1$. Then show that $n!$ has length at least equal to $\frac{n}{2}(\log_2 n - 2)$, and that for large n this is greater than $C'n \ln n$ for a suitable constant C'.

5. Use Stirling's formula to find a simple function $g(n)$ such that the length of $n!$ is asymptotically equal to $g(n)$.

6. Suppose that the letters $A, B, C, \ldots, Z$ are used as base-26 digits. Then the binary length of

$$(DOG)_{26}^{(CAT)_{26}}$$

is roughly equal to (choose one):

$50, 150, 500, 1500, 5000, 15000, 50000, 150000, 500000, 1500000, 5000000$.

7. Arrange the following numbers in increasing order, if n is equal to the U.S. national debt measured in kopecks:*
  (a) The number of decimal digits in 2^n.
  (b) The number of consecutive zeros at the end of the binary expansion of $n!$.
  (c) The binary length of the value at n of a quintic polynomial whose coefficients are 20-bit integers.
  (d) The binary length of $[\sqrt{n}]!$.
  (e) The number of primes you have to try to divide into n if you want to be sure that n is a prime number using the method of trial division.

§ 3. Time Estimates

As mentioned before, we shall assume that all arithmetic is being done in binary, i.e., with 0's and 1's.

3.1 Bit Operations

Let us start with a very simple arithmetic problem, the addition of two binary integers, for example:

$$\begin{array}{r} {\scriptstyle 1\,111} \\ 1111000 \\ + \underline{0011110} \\ 10010110 \end{array}$$

Suppose that the numbers are both k bits long; if one of the two integers has fewer bits than the other, we fill in zeros to the left, as in this example, to make them have the same length. Although this example involves small integers (with $k = 7$), we should think of k as perhaps being very large, like 500 or 1000.

Let us analyze in complete detail what this addition entails. Basically, we must repeat the following steps k times:

1. Look at the top and bottom bit and also at whether there's a carry above the top bit.

---

* Currently the U.S. national debt is about $\$5 \times 10^{12}$, and one dollar is worth approximately 5000 rubles. (And one ruble is 100 kopecks.)

2. If both bits are 0 and there is no carry, then put down 0 and move on.

3. If either (a) both bits are 0 and there is a carry, or (b) one of the bits is 0, the other is 1, and there is no carry, then put down 1 and move on.

4. If either (a) one of the bits is 0, the other is 1, and there is a carry, or else (b) both bits are 1 and there is no carry, then put down 0, put a carry in the next column, and move on.

5. If both bits are 1 and there is a carry, then put down 1, put a carry in the next column, and move on.

Doing this procedure once is called a *bit operation*. Adding two k-bit numbers requires k bit operations. We shall see that more complicated tasks can also be broken down into bit operations. The amount of time a computer takes to perform a task is essentially proportional to the number of bit operations. Of course, the constant of proportionality – the fraction of a nanosecond per bit operation – depends on the particular computer system. (This is an over-simplification, since the time can be affected by "administrative matters", such as accessing memory.) When we speak of estimating the "time" it takes to accomplish something, we mean finding an estimate for the number of bit operations required.

Thus, the time required (i.e., number of bit operations) to add two numbers is equal to the *maximum of the lengths of the two numbers*. We write:

$$\text{Time}(k\text{-bit} + l\text{-bit}) = \max(k, l) \ .$$

If we want to express the time in terms of the two numbers added, say m and n, then, since $k = \text{length}(m) = O(\ln m)$, we have

$$\text{Time}(m + n) = O\big(\max(\ln m, \ln n)\big) \ .$$

Notice that there's a big difference between expressing the time for performing a task on some integers in terms of the *integers themselves* (in this case m and n) and in terms of the *lengths of the integers* (in this case k and l). Depending on the situation, either type of time estimate might be convenient for us to use. It's important not to confuse them.

Next, let's examine the process of *multiplying* a k-bit integer by an l-bit integer in binary. For example,

$$
\begin{array}{r}
11101 \\
\underline{1101} \\
11101 \\
111010 \\
\underline{11101} \\
101111001
\end{array}
$$

In general, suppose that we use this familiar procedure to multiply a k-bit integer n by an l-bit integer m. We obtain at most l rows (one row fewer for each 0 bit in m), where each row consists of a copy of n shifted to the left a certain distance – that is, with zeros put on at the right end. In order to count bit operations, we suppose that we perform the addition two rows at a time, by first adding the second row to the first, then adding the third row to the result from the first addition, then adding the fourth row to the result of the second addition, and so on. In other words, we need to perform at most $l - 1$ additions. In each addition we first copy down the right-most bits from the top row that are above the places in the lower row where we filled in zeros. This process of simply transfering the bits down counts as an "administrative procedure", not as bit operations, and so is neglected in our time estimate. So each addition requires only k bit operations. Thus, the total number of bit operations to get our answer is less than

$$(l \text{ additions}) \times (k \text{ bit operations per addition}) = kl .$$

Before giving other examples of time estimates, we should make several observations. In the first place, we define the time it takes to perform an arithmetic task to be an upper bound for the number of bit operations, without including any consideration of shift operations, memory access, etc.

In the second place, if we want to get a time estimate that is simple and convenient to work with, we should assume at various points that we're in the "worst possible case". For example, in a multiplication we might have a lot fewer than $l - 1$ additions of nonzero rows. But if we are interested only in big-O estimates, then we get no benefit by taking this into account.

Time estimates do not have a single "right answer". For example, regarding the time required to multiply a k-bit number by an l-bit number, all of the following statements are correct: (1) Time = $O(kl)$; (2) Time < kl; (3) Time $\leq k(l - 1)$; (4) if the second number has an equal number of 0-bits and 1-bits, then Time $\leq kl/2$. In what follows, we shall always use either the estimate Time < kl or else Time = $O(kl)$.

Next, we note that our time estimate can be expressed in terms of the numbers multiplied rather than their lengths, as follows:

$$\text{Time}(m \times n) = O(\ln m \ln n) .$$

As a special case, if we want to multiply two numbers of about the same size, we can use the estimate

$$\text{Time}(k\text{-bit} \times k\text{-bit}) = O(k^2) .$$

It should be noted that much work has been done on increasing the speed of multiplying two k-bit integers when k is large. With the help of techniques that are much more complicated than the grade-school method we have been using, mathematicians have been able to find a procedure for multiplying two k-bit integers that requires only $O(k \ln k \, \ln \ln k)$ bit operations. This is better than $O(k^2)$, and even better than $O(k^{1+\varepsilon})$ for any $\varepsilon > 0$, no matter how small. However, in what follows we shall always be content to use the weaker estimates above for the time needed for a multiplication.

3.2 Algorithms

In general, when estimating the number of bit operations required to do something, the first step is to decide upon and write an outline of a detailed procedure for performing the task. We did this earlier in the case of our multiplication problem. An explicit step-by-step procedure for doing calculations is called an *algorithm*. Of course, there may be many different algorithms for doing the same thing. One may choose to use the easiest one to write down, or one may choose to use the fastest one known, or one may choose to compromise and make a trade-off between simplicity and speed. The algorithm used above for multiplying n by m is far from the fastest one known. But it is certainly a lot faster than repeated addition (adding n to itself m times).

So far we have discussed addition and multiplication in binary. Subtraction works very much like addition: we have the same estimate $O(k)$ for the amount of time required to subtract two k-bit integers. However, we have to slightly broaden the definition of bit operation to include subtraction. That is, a subtraction bit operation can be defined just as the addition bit operation was before, except with "borrows" instead of "carries" and a different list of four alternatives.

Division can be analyzed in much the same way as multiplication, with the result that it takes $O\big(l(k-l+1)\big)$ bit operations to obtain the quotient and remainder when a k-bit integer is divided by an l-bit integer, where $k \geq l$ (of course, if $k < l$, then the quotient is 0 and the "division" is trivial). In other words, given two positive integers $b < a$, the time it takes to find q and r such that $a = qb + r$, where $0 \leq r < b$, depends on the product of the lengths of b and q. (See Exercise 1 of §2.)

Example 3.1. Estimate the time required to convert a k-bit integer n to its representation in the base 10.

Solution. The conversion algorithm is as follows. Divide $10 = (1010)_2$ into n. The remainder – which will be one of the integers 0, 1, 10, 11, 100, 101, 110, 111, 1000, or 1001 – will be the ones digit d_0. Now replace n by the quotient and repeat the process, dividing that quotient by $(1010)_2$, using the remainder as d_1 and the quotient as the next number into which to divide $(1010)_2$. This process must be repeated a number of times equal to the number of decimal digits in n, which is $\left[\frac{\ln n}{\ln 10}\right] + 1 = O(k)$. Then we're done. (We might want to take our list of decimal digits, i.e., of remainders from all the divisions, and convert them to the more familiar notation by replacing 0, 1, 10, 11, ..., 1001 by 0, 1, 2, 3, ..., 9, respectively.) How many bit operations does this all take? Well, we have $O(k)$ divisions, each requiring $O(4k)$ operations (dividing a number with at most k bits by the 4-bit number $(1010)_2$). But $O(4k)$ is the same as $O(k)$ (constant factors don't matter in the big-O notation), so we conclude that the total number of bit operations is $O(k) \cdot O(k) = O(k^2)$. If we want to express this in terms of n rather than k, then since $k = O(\ln n)$, we can write

$$\text{Time(convert } n \text{ to decimal)} = O(\ln^2 n) \ .$$

Example 3.2. Estimate the time required to convert a k-bit integer n to its representation in the base b, where b might be very large.

Solution. Using the same algorithm as in Example 3.1, except dividing now by the l-bit integer b, we find that each division takes longer than before (if l is large), namely, $O(kl)$ bit operations. How many times do we have to divide? Here notice that the number of base-b digits in n is $O(k/l)$. Thus, the total number of bit operations required to do all of the necessary divisions is $O(k/l) \cdot O(kl) = O(k^2)$. This turns out to be the same answer as in Example 3.1. That is, our estimate for the conversion time does not depend upon the base to which we're converting (no matter how large it may be). This is because the greater time required to find each digit is offset by the fact that there are fewer digits to be found.

Example 3.3. Estimate the time required to compute $n!$.

Solution. We use the following algorithm. First multiply 2 by 3, then the result by 4, then the result of that by 5,..., until you get to n. At the $(j - 1)$-th step you're multiplying $j!$ by $j + 1$. Here you have $n - 2$ multiplications, where each multiplication involves multiplying a partial product (namely, $j!$) by the next integer. The partial product will start to be very large. As a worst case estimate for the number of bits it has, let's take the number of binary digits in the last product, namely, in $n!$. According to Example 2.2, length$(n!) = O(n \ln n)$.

Thus, in each of the $n - 2$ multiplications in the computation of $n!$, we are multiplying an integer with at most $O(\ln n)$ bits (namely, $j + 1$) by an integer with $O(n \ln n)$ bits (namely, $j!$). This requires $O(n \ln^2 n)$ bit operations. We must do this $n - 2 = O(n)$ times. So the total number of bit operations is $O(n \ln^2 n) \cdot O(n) = O(n^2 \ln^2 n)$. We end up with the estimate: Time(computing $n!$) $= O(n^2 \ln^2 n)$.

3.3 The Euclidean Algorithm

Example 3.4. Show that, given two integers $a > b > 0$, the greatest common divisor* d of a and b can be computed and the equation

$$au + bv = d$$

can be solved for integers u and v in time $O(\ln a \ln b)$.

Solution. We recall the extended Euclidean algorithm. First, we successively divide

---

  * Recall from elementary number theory that the *greatest common divisor* of a and b, abbreviated g.c.d., is the largest positive integer d that divides both a and b; if this integer is 1, then a and b are said to be *relatively prime*.

$$a = q_0 b + r_1 , \qquad 0 < r_1 < b ,$$
$$b = q_1 r_1 + r_2 , \qquad 0 < r_2 < r_1 ,$$
$$r_1 = q_2 r_2 + r_3 , \qquad 0 < r_3 < r_2 ,$$
$$\vdots \qquad \vdots$$
$$r_{j-1} = q_j r_j + r_{j+1} , \qquad 0 < r_{j+1} < r_j ,$$
$$\vdots \qquad \vdots$$
$$r_{l-2} = q_{l-1} r_{l-1} + r_l , \qquad 0 < r_l < r_{l-1} ,$$
$$r_{l-1} = q_l r_l + r_{l+1} , \qquad 0 < r_{l+1} < r_l ,$$
$$r_l = q_{l+1} r_{l+1} ,$$

so that $d = r_{l+1}$. Then we work backwards, writing

$$d = r_{l+1} = r_{l-1} - q_l r_l$$
$$= v_l r_l + u_{l-1} r_{l-1} , \qquad v_l = -q_l , \quad u_{l-1} = 1$$
$$= v_{l-1} r_{l-1} + u_{l-2} r_{l-2} , \qquad v_{l-1} = u_{l-1} - q_{l-1} v_l , \quad u_{l-2} = v_l ,$$
$$\vdots$$
$$= v_j r_j + u_{j-1} r_{j-1} , \qquad v_j = u_j - q_j v_{j+1} , \quad u_{j-1} = v_{j+1} ,$$
$$\vdots$$
$$= v_1 r_1 + u_0 b , \qquad v_1 = u_1 - q_1 v_2 , \quad u_0 = v_2 ,$$
$$= vb + ua , \qquad v = u_0 - q_0 v_1 , \quad u = v_1 .$$

To estimate the time required for all this, we recall that the number of bit operations in the division $a = q_0 b + r_1$ is at most length(b) · length(q_0). Similarly, the time for the division $r_{j-1} = q_j r_j + r_{j+1}$ is at most length(r_j) · length(q_j) ≤ length(b) · length(q_j). Thus, the total time for all the divisions is $O\big(\ln b(\ln q_0 + \ln q_1 + \cdots + \ln q_{l+1})\big) = O\big((\ln b)(\ln \prod q_j)\big)$. But it is easy to show by induction that $\prod q_j \le a$, and so the bound is $O(\ln b \ln a)$. We leave it to the reader to show that the number of bit operations required to "work backwards" in the Euclidean algorithm – that is, to compute all of the $v_j = u_j - q_j v_{j+1}$, is also $O(\ln b \ln a)$. Thus, the extended Euclidean algorithm takes time $O(\ln^2 a)$.

An immediate consequence of Example 3.4 is that any congruence

$$ax \equiv 1 \pmod{m}$$

with $|a| < m$ and g.c.d.$(a, m) = 1$ can be solved for x in time $O(\ln^2 m)$.

Example 3.5. Suppose that m is a k-bit natural number, N is an l-bit natural number, and $|b| < m$. Show how to find the least nonnegative residue of b^N modulo m in time $O(k^2 l)$.

Solution. This is done by the "repeated squaring method" of modular exponentiation (also called the "square and multiply" method). We first write N in binary:

$N = \varepsilon_{l-1} \cdot 2^{l-1} + \varepsilon_{l-2} \cdot 2^{l-2} + \cdots + \varepsilon_1 \cdot 2 + \varepsilon_0$. We then successively compute the least nonnegative residue of b^{2^j} modulo m for $j = 1, \ldots, l - 1$. To compute b^{2^j} we take the value just computed for $b^{2^{j-1}}$ modulo m, square it, and reduce modulo m. Since none of the numbers we work with could have length more than $2k$ (because multiplying two residues in $\{0, 1, \ldots, m - 1\}$ gives a number less than m^2), this process takes time $O(k^2)$ for each j.

Next, let $j_1, j_2, \ldots, j_\lambda$ be the indices for which $\varepsilon_{j_\nu} = 1$, i.e., the locations of all 1-bits in N. Then $N = \sum 2^{j_\nu}$ and $b^N = \prod b^{2^{j_\nu}}$. We first multiply the least nonnegative residue of $b^{2^{j_1}}$ and the least nonnegative residue of $b^{2^{j_2}}$, and reduce the result modulo m; then we multiply this result by the least nonnegative residue of $b^{2^{j_3}}$ and reduce modulo m; and so on. The final result will be b^N. It is clear that the time required for the repeated squaring algorithm is $O(k^2 l)$.

3.4 From Polynomial Time to Exponential Time

We now make a definition that is fundamental in the study of algorithms.

Definition 3.1. An algorithm to perform a computation is said to be a *polynomial time* algorithm if there exists an integer d such that the number of bit operations required to perform the algorithm on integers of total length at most k is $O(k^d)$.

Thus, the usual arithmetic operations $+$, $-$, $\times$, $\div$ are examples of polynomial time algorithms; so is conversion from one base to another. On the other hand, computation of $n!$ is not. (However, if one is satisfied with knowing $n!$ to only a certain number of significant figures, e.g., its first 1000 binary digits, then one can obtain that by a polynomial time algorithm using Stirling's approximation formula for $n!$.)

Remark. The words "perform the algorithm on integers of total length at most k" in Definition 3.1 are a little vague. What is meant is the following. When we set up a computation, strictly speaking, we should always specify the form of the "input". Then k in Definition 3.1 stands for the total binary length of the input. In many problems the form of the input is obvious, and is usually not stated explicitly. In Examples 3.2 and 3.3, the input was the number n written in binary. However, sometimes one has to be careful, as the following example shows.

Example 3.6. Is there a polynomial time algorithm for determining whether the m-th Fermat number is prime or composite?

Here it is crucial to specify the form of the input. If the input is the number $n = 2^{2^m} + 1$ written in binary (i.e., $100 \cdots 001$ with $2^m - 1$ zeros between the two 1's), then the answer to this question is "yes". That is, there are several algorithms that can determine whether n is prime or composite in time that is bounded by a polynomial function of 2^m. However, if the input is the number m written in binary, then the answer to this question is almost certainly "no". There is no known algorithm that can determine primality of the m-th Fermat number in time that is bounded by a fixed power of $\log_2 m$.

One class of algorithms that are very far from polynomial time is the class of *exponential time* algorithms. These have a time estimate of the form $O(e^{ck})$, where c is a constant. Here k is the total binary length of the integers to which the algorithm is being applied. For example, the "trial division" algorithm for factoring an integer n can easily be shown to take time $O(n^{1/2+\varepsilon})$ (where $\varepsilon > 0$ can be arbitrarily small). Since $k \approx \log_2 n$, the expression inside the big-O can also be written as e^{ck}, where $c = (\frac{1}{2} + \varepsilon) \ln 2$.

There is a useful way to classify time estimates in the range between polynomial and exponential time. Let n be a large positive integer, perhaps the input for our algorithm; let γ be a real number between 0 and 1; and let $c > 0$ be a constant.

Definition 3.2. Let

$$L_n(\gamma; c) = O\left(e^{c\left((\ln n)^\gamma (\ln \ln n)^{1-\gamma}\right)}\right) .$$

In particular, $L_n(1; c) = O(e^{c \ln n}) = O(n^c)$, and $L_n(0; c) = O(e^{c \ln \ln n}) = O((\ln n)^c)$. An $L(\gamma)$-*algorithm* is an algorithm that, when applied to the integer n, has running time estimate of the form $L_n(\gamma; c)$ for some c. In particular, a *polynomial time* algorithm is an $L(0)$-algorithm, and an *exponential time* algorithm is an $L(1)$-algorithm. By a *subexponential time* algorithm we mean an $L(\gamma)$-algorithm for some $\gamma < 1$.

Roughly speaking, γ measures the fraction of the way we are from polynomial to exponential time. We saw that the naive trial division algorithm to factor n is an $L(1)$-algorithm. Until recently, the best general (probabilistic) factoring algorithms were all $L(1/2)$-algorithms. Then with the advent of the "number field sieve" (see [Lenstra and Lenstra 1993]) the difficulty of factoring was pushed down to $L(1/3)$.

The $L(\gamma)$-terminology is not appropriate for all algorithms. For example, algorithms that take much more than exponential time cannot be classified by Definition 3.2. Nor is the $L(\gamma)$-terminology useful for algorithms that are just slightly slower than polynomial time – such as the $O\left((\ln n)^{c \ln \ln \ln n}\right)$ primality test in [Adleman, Pomerance, and Rumely 1983].

Some people prefer to give a different definition of "subexponential time". They use the term for an algorithm with running time bounded by a function of the form $e^{f(k)}$, where k is the input length and $f(k) = o(k)$ (see Remark 6 of §1 for the meaning of little-o). For example, an algorithm taking $e^{k/\ln \ln k}$ operations would be subexponential time in this sense, but not in the sense of Definition 3.2.

Exercises for § 3

1. (a) Using the big-O notation, estimate in terms of a simple function of n the number of bit operations required to compute 3^n in binary.
(b) Do the same for n^n.

(c) Estimate in terms of a simple function of n and N the number of bit operations required to compute N^n.

2. The number of bit operations required to compute the exact value of

$$10110111001011^{1000111}$$

(where the numbers are written in binary) is roughly equal to (choose one):

$$100, \ 1000, \ 10000, \ 100000, \ 1000000, \ 10^{10}, \ 10^{25}, \ 10^{75} \ .$$

3. The following formula holds for the sum of the first n perfect squares:

$$\sum_{j=1}^{n} j^2 = n(n+1)(2n+1)/6 \ .$$

(a) Using the big-O notation, estimate (in terms of n) the number of bit operations required to perform the computations in the left side of this equality.
(b) Estimate the number of bit operations required to perform the computations on the right in this equality.

4. Suppose that you have an algorithm that solves a problem whose input is a single integer. Let k denote the binary length of this integer. You are interested in applying this algorithm to numbers of binary length about $k = 1000$. You test the algorithm on numbers of length about 100, and find that your computer takes about 1 minute to carry out the algorithm for each such number. How much time will your computer take to apply the algorithm to a number of binary length $k = 1000$ if the time estimate for the algorithm is

(a) Ck^3 bit operations, where C is some constant?

(b) $Ce^{0.03k}$ bit operations, where C is some constant?

In each case choose your answer from among the following: (A) 10 minutes; (B) 100 minutes; (C) 16 hours; (D) 1 week; (E) 2 months; (F) 2 years; (G) 100 years; (H) 10000 years; (I) 1000000 years; (J) not enough information given to answer the question.

5. (a) Using the big-O notation, estimate the number of bit operations required to find the **sum** of the first n Fibonacci numbers (see Exercise 2(f) of §2).

(b) The same for their **product**.

6. Suppose that you have a list of all primes having k or fewer bits. Using the Prime Number Theorem and the big-O notation, estimate the number of bit operations needed to compute
(a) the sum of all of these primes;
(b) the product of all of these primes;
(c) the k most significant bits in the product of all of these primes.

7. Suppose that m is a k-bit integer, and n is an l-bit integer (and you don't know in advance whether k is much bigger than l, l is much bigger than k, or they're

about the same size). Find a bound of the form $O(g(k, l))$ for the number of bit operations required to compute $m^3 n^4$. Your function $g(k, l)$ should be as simple and efficient as possible.

8. Given a k-bit integer, you want to compute the highest power of this number that has l or fewer bits. (Suppose that l is much larger than k.) Estimate the number of bit operations required to do this. Your answer should be a very simple expression in terms of k and/or l.

9. Suppose that we are given l different moduli m_i such that g.c.d.$(m_i, m_j) = 1$ for $i \neq j$, and l integers a_i such that $|a_i| < m_i$. Let $M = \prod_i m_i$. According to the Chinese Remainder Theorem, there exists a unique x in the range $0 \leq x < M$ such that $x \equiv a_i \pmod{m_i}$ for $i = 1, \ldots, l$. Suppose that all of the moduli m_i are k-bit integers. In parts (a)–(g) below we recall the steps in the algorithm for finding x. For each step find a big-O estimate in terms of k and l for the number of bit operations required.
(a) Compute M.
(b) For each i compute $M_i = M/m_i$.
(c) For each i find the least positive residue of M_i modulo m_i.
(d) For each i find the least positive y_i that satisfies $y_i M_i \equiv 1 \pmod{m_i}$.
(e) For each i compute $a_i M_i y_i$.
(f) Add all of the numbers in part (e).
(g) Find the least nonnegative residue modulo M of the number in part (f). This is the desired value x.
(h) Let K denote the total length of the input (i.e., the l-tuple of a_i and the l-tuple of m_i). Note that $kl < K \leq 2kl$. Find a big-O bound in terms of K for the number of bit operations required to go through all of the steps in the above Chinese Remainder Theorem algorithm.

10. Arrange the following numbers in increasing order, if n is equal to the number of mosquitos in New Jersey:
(a) the time required to solve a Chinese Remainder Theorem problem with approximately $\ln n$ congruences whose moduli satisfy $n < m_i < 2n$.
(b) the time required to find the value at n of a quintic polynomial whose coefficients are 20-bit integers;
(c) the time required to convert n (which is initially written in binary) to hexadecimal (base 16);
(d) the time required to find the least nonnegative residue of $m!$ modulo p, where m is an integer of approximately the same size as $\ln n$ and p is a prime of approximately the same size as $2 \ln n$;
(e) the time required to compute the least nonnegative residue of b^n modulo m, where b and m are numbers of approximately the same size as n.

11. Suppose that an algorithm requires $L_n(\gamma; 1)$ microseconds when applied to the integer n (where the constant in the big-O in Definition 3.2 is taken to be 1). Find the time required to apply the algorithm to a number $n \approx 10^{100}$ when $\gamma = 0, 1/3, 1/2$, and 1.

§ 4. P, NP, and NP-Completeness

This section is devoted to three fundamental notions of computer science: the class P of decision problems solvable in polynomial time, the class NP of decision problems solvable in nondeterministic polynomial time, and the class of NP problems that are "complete". We shall give only an informal introduction to P, NP, and NP-completeness. For greater rigor and more details, the reader is referred to standard books on complexity theory, such as [Garey and Johnson 1979] and [Papadimitriou 1994].

4.1 Problem Instances, Search Problems, and Decision Problems

In what follows, the term "problem" refers to a general description of a task, and the term "instance" of a problem means a particular case of the task.

Example 4.1. The Integer Factorization search problem is the problem of either finding a nontrivial factor M of an integer N or else determining that no nontrivial factor exists – in other words, that N is prime. Once we are given a particular value of N and are asked to factor it, we have an *instance* of the Integer Factorization search problem.

Example 4.2. The Traveling Salesrep problem is the task of finding the shortest route that starts from City A, passes through all other cities on the salesrep's list, and returns to City A. An *instance* of the Traveling Salesrep problem is a specific list of cities and the distances between any pair of cities. (Depending on what it is that the salesrep wants to minimize, instead of distances she might have a list of the airfare between any two cities or the total cost of travel between the two cities.)

Example 4.3. The 3-Coloring problem is the task of coloring a given map with just three colors in such a way that no two neighboring regions have the same color, if it is possible to do so. Actually, it is more natural to study the problem of coloring a graph rather than a map, because that is more general (see Exercise 4 below). To be precise, a "graph" is a list of dots (called "vertices") and lines (called "edges") joining certain pairs of dots. The 3-Coloring problem for graphs is the task of assigning one of three colors to each vertex in such a way that no two vertices that are joined by an edge have the same color.

An example of a 3-colorable graph is shown at the top of the next page.

The term "input" refers to all the information that must be specified in order to describe an instance of the problem. In Integer Factorization the input is simply the integer N. The term "input length" refers to the number of symbols needed to list the input. We suppose that a particular system of symbols has been fixed once and for all. For example, if we are writing numbers in binary, then the input length for an instance of Integer Factorization is $1 + [\log_2 N]$.

In the Traveling Salesrep problem with m cities, if we suppose that the cities are numbered from 1 to m, then the input is a particular map from the set of pairs

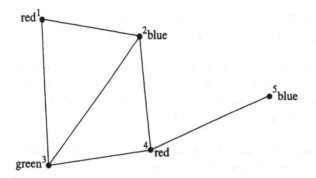

(i, j), $1 \leq i < j \leq m$, to the set of natural numbers $\mathbb{N}$. (We are supposing that all of the distances are positive integers.)

In a 3-Coloring problem with m vertices, if we suppose that the vertices are labeled from 1 to m, the input may be regarded as a subset of the set of pairs (i, j), $1 \leq i < j \leq m$. That is, the input is a graph $G = (V, E)$, where V is the vertex set $\{1, \ldots, m\}$ and $E \subset \{(i, j)\}_{1 \leq i < j \leq m}$ is the set of edges.

In order to give the definitions of P and NP, we first have to modify our problems so that they are "decision problems". A decision problem is a problem whose solution (output) consists of a yes-or-no answer. On the other hand, if the desired output is more than a "yes" or "no" – that is, if we want to find a number, a route on a map, etc. – then we call the problem a "search problem".

Remark. Unlike a decision problem, a search problem might have several correct answers. For example, in the Traveling Salesrep search problem we want a path of minimal length that passes through all the cities. (A path passing through all the cities and returning to its starting point is sometimes called a "tour".) There may be many different minimal tours.

Example 4.4. An instance of a decision problem version of Integer Factorization is as follows:

INPUT: Positive integers N and k.
QUESTION: Does N have a factor M satisfying $2 \leq M \leq k$?

The problem of actually finding a nontrivial factor M of N is called the Integer Factorization search problem.

Example 4.5. An instance of the Traveling Salesrep decision problem has the form

INPUT: An integer m, a map from the set of pairs (i, j), $1 \leq i < j \leq m$, to the natural numbers, and an integer k.
QUESTION: Is there a tour of the cities of length $\leq k$?

The Traveling Salesrep search problem is the problem of finding a tour of minimal length.

Example 4.6. An instance of the 3-Coloring decision problem has the form

INPUT: A graph $G = (V, E)$.

QUESTION: Does this graph have a 3-coloring? In other words, does there exist a map c from V to a 3-element set such that $(i, j) \in E \Longrightarrow c(i) \neq c(j)$?

For many problems – including Integer Factorization, Traveling Salesrep, and 3-Coloring – the decision problem and the search problem are essentially equivalent. This means that an algorithm to do one can easily be converted into an algorithm to do the other. Let us see how this works in the case of Integer Factorization.

First, suppose that we have an algorithm to do the search problem. This means that, given N, we can apply the algorithm to find a nontrivial factor M, then apply it again to find nontrivial factors of M and N/M, and so on, until N has been written as a product of prime powers. Once we have the prime factorization of N, we can immediately determine whether or not N has a factor in the interval $[2, k]$. Namely, the answer to this question is "yes" if and only if the smallest prime divisor of N is in that interval.

Conversely, suppose that we have an algorithm to do the decision problem. In that case we can use the method of "20 questions" (also called *binary search*) in order to zero in on the exact value of a factor, thereby solving the Integer Factorization search problem. More precisely, we find a nontrivial factor of N bit by bit, starting with its leading bit. Let 2^n be the smallest power of 2 that is larger than N. In other words, n is the input length $1 + [\log_2 N]$. First we apply the decision problem algorithm with $k = 2^{n-1} - 1$. If the answer is "no", then N is prime, because any nontrivial factor M must satisfy $M \leq N/2 < 2^{n-1}$. In that case we're done. Now suppose that the answer is "yes". Repeat the algorithm for the decision problem with $k = 2^{n-2} - 1$. If the answer is "no", then N must have a nontrivial factor of the form $M = 1 \cdot 2^{n-2} + \varepsilon_{n-3} 2^{n-3} + \cdots + \varepsilon_0$, where the ε_i are the bits in the binary representation of M. If the answer is "yes", then N must have a nontrivial factor of the same form but with first bit zero rather than one, i.e., $M = \varepsilon_{n-3} 2^{n-3} + \cdots + \varepsilon_0$. To find the next bit ε_{n-3}, either set $k = 2^{n-2} + 2^{n-3} - 1$ (in the case when the previous application of the algorithm gave a "no" answer) or else set $k = 2^{n-3} - 1$ (in the case when the previous application of the algorithm gave a "yes" answer). If the algorithm now answers "no", then you know that you should choose $\varepsilon_{n-3} = 1$; if it answers "yes", then you may choose $\varepsilon_{n-3} = 0$. Continue in this manner, applying the algorithm for the decision problem once to find each bit in a factor of N. After only n applications of the algorithm, you will have found a nontrivial factor of N. So the algorithm for the decision problem has been converted into an algorithm for the corresponding search problem.

Example 4.7. Given an algorithm for the Integer Factorization decision problem, here is how we use it to factor 91.

First question to algorithm: Does 91 have a factor between 2 and 63? Answer: YES.

Second question: Does 91 have a factor between 2 and 31? Answer: YES.
Third question: Does 91 have a factor between 2 and 15? Answer: YES.
Fourth question: Does 91 have a factor between 2 and 7? Answer: YES.

Fifth question: Does 91 have a factor between 2 and 3? Answer: NO.
Sixth question: Does 91 have a factor between 2 and 5? Answer: NO.
Seventh question: Does 91 have a factor between 2 and 6? Answer: NO.
We conclude that 7 is a nontrivial factor of 91. Note that this method of binary search using an algorithm for the Integer Factorization decision problem will always lead to the smallest nontrivial factor of N.

Thus, from the standpoint of computer science, there is often no loss of generality in working with decision problems rather than search problems.

4.2 P and NP

Definition 4.1. A decision problem $\mathcal{P}$ is in the class P of *polynomial time* problems if there exists a polynomial $p(n)$ and an algorithm such that if an instance of $\mathcal{P}$ has input length $\leq n$, then the algorithm answers the question correctly in time $\leq p(n)$.* An equivalent definition is: A decision problem $\mathcal{P}$ is in P if there exists a constant c and an algorithm such that if an instance of $\mathcal{P}$ has input length $\leq n$, then the algorithm answers the question in time $O(n^c)$.

Notice the close relation between Definitions 4.1 and 3.1: a decision problem is in P if there exists a polynomial time algorithm (in the sense of Definition 3.1) that solves it.

Remark. Definition 4.1 attempts to capture a class of problems that in practice can be solved rapidly. It is not *a priori* clear that P is the right class to take for this purpose. For instance, an algorithm with running time n^{100}, where n is the input length, is slower than one with running time $e^{0.0001n}$ until n is greater than about ten million, even though the first algorithm is polynomial time and the second one is exponential time. In this connection see §7.2 of Chapter 1.

However, the experience has been that if a problem of practical interest is in P, then there is an algorithm for it whose running time is bounded by a *small* power of the input length. Sometimes a problem that is in P or is believed to be in P has a practical, efficient algorithm that is not polynomial time. An example is the following Primality problem:

INPUT: A positive integer N.
QUESTION: Is N a prime number?

If the so-called "Extended Riemann Hypothesis" is true, then an algorithm in [Miller 1976] will answer this question in polynomial time. However, even if one assumes the ERH, for $N < 10^{1000}$ the most efficient deterministic** algorithm

---

* We suppose that we have a fixed computer to implement the algorithm, and "time" refers to the running time on this computer. Alternatively, we could define "time" to be the number of bit operations required to carry out the algorithm.
** All of the algorithms discussed so far in this chapter are deterministic; the term "deterministic" is used to distinguish these algorithms from "randomized" (also called "probabilistic") algorithms (see §6).

known is the method using Gauss and Jacobi sums (see [Adleman, Pomerance, and Rumely 1983] and [Cohen and Lenstra 1984]), which has running time $n^{O(\ln \ln n)}$, where $n = O(\ln N)$ is the input length.

An example of a slightly different sort is given by the problem

INPUT: An elliptic curve E modulo p (see Chapter 6), and an integer k.
QUESTION: Are there $\geq k$ points on E?

The algorithm in [Schoof 1985] answers this question in time $O(n^8)$, where $n = O(\ln p)$ is the input length. There is an algorithm due to Atkin that is much more efficient in practice, but no one can prove a rigorous bound on its running time; in particular, Atkin's algorithm is not known to be polynomial time.

Thus, empirically it seems that the problems in P that are of practical interest all have efficient algorithms, although in some cases the most efficient algorithms are different from the polynomial time algorithms and in other cases they are not the ones that lend themselves to a rigorous analysis of the running time.

Definition 4.2. A decision problem $\mathcal{P}$ is in the class NP if, given any instance of $\mathcal{P}$, a person with unlimited computing power not only can answer the question, but in the case that the answer is "yes", she can supply evidence that another person could use to verify the correctness of the answer in polynomial time. Her demonstration that her "yes" answer is correct is called a "certificate" (more precisely, a *polynomial time certificate*).

A decision problem $\mathcal{P}$ is said to be in the class co-NP if the above condition holds with "yes" replaced by "no". That is, for any instance having a "no" answer there must exist a polynomial time certificate that the "no" answer is correct.

Example 4.8. Consider the above decision version of Integer Factorization:

INPUT: Positive integers N and k.
QUESTION: Does N have a factor in the interval $[2, k]$?

This problem is almost certainly not in P. However, it is in NP. Namely, suppose that an all-powerful person (such as God or an advanced extraterrestrial being) factors N and finds that it has a factor $M \in [2, k]$. After this person tells you that the answer is "yes", she supplies M, from which you can verify the correctness of her answer in polynomial time, simply by dividing M into N.

The Integer Factorization problem is also in co-NP, although to see this requires more thought. If the answer to the above question is "no", the extraterrestrial gives you the complete prime factorization of N, from which you can immediately see that there is no prime factor $\leq k$. At the same time she must also give you a certificate with which you can verify in polynomial time that each of the prime factors really is prime. There are various certificates that can be given; the oldest and simplest is due to Pratt (see, for example, Theorem 8.20 of [Rosen 1993]).

Example 4.9. The Traveling Salesrep decision problem is almost certainly not in P (we'll have more to say about that later). However, it is in NP. That is, suppose that an extraterrestrial being finds the most economical tour for the traveling salesrep, and it turns out to have length less than or equal to k. She tells you that the answer is "yes", and then shows you the route, at which point you can rapidly verify that her "yes" answer is in fact correct.

In the same way, one easily sees that the 3-Coloring decision problem is in NP.

If a problem is in P, then trivially it is in NP. That is, P⊂NP. It is almost certain that NP is a much bigger class of problems than P, but this has not been proved. The claim that P≠NP is the most famous conjecture in computer science.

4.3 Reducing One Problem to Another

Definition 4.3. Let $\mathcal{P}_1$ and $\mathcal{P}_2$ be two decision problems. We say that $\mathcal{P}_1$ *reduces to* $\mathcal{P}_2$ (more precisely, *reduces to* $\mathcal{P}_2$ *in polynomial time*) if there exists an algorithm that is polynomial time as a function of the input length of $\mathcal{P}_1$ and that, given any instance P_1 of $\mathcal{P}_1$, constructs an instance P_2 of $\mathcal{P}_2$ such that the answer for P_1 is the same as the answer for P_2.

One basic use for this notion of reduction is as follows. Suppose that we have an efficient algorithm for $\mathcal{P}_2$. If $\mathcal{P}_1$ reduces to $\mathcal{P}_2$, then we can use the algorithm for $\mathcal{P}_2$ to solve $\mathcal{P}_1$ as well. Namely, given an instance of $\mathcal{P}_1$, in polynomial time we find a corresponding instance of $\mathcal{P}_2$ using the algorithm in Definition 4.3. Then if we apply our algorithm for $\mathcal{P}_2$ to this instance of $\mathcal{P}_2$, the answer we get is also the answer to our original $\mathcal{P}_1$ question. That is, an algorithm for $\mathcal{P}_2$ automatically gives an algorithm for $\mathcal{P}_1$. If our algorithm for $\mathcal{P}_2$ is a polynomial time algorithm, then so is the resulting algorithm for $\mathcal{P}_1$.

Example 4.10. Let $\mathcal{P}_1$ be the following problem:

INPUT: A quadratic polynomial $p(X)$ with integer coefficients.
QUESTION: Does $p(X)$ have two distinct real roots?

Let $\mathcal{P}_2$ be the problem

INPUT: An integer N.
QUESTION: Is N positive?

We show that $\mathcal{P}_1$ reduces to $\mathcal{P}_2$. Let $p(X) = aX^2 + bX + c$ be an instance of $\mathcal{P}_1$. Set $N = b^2 - 4ac$, which can be computed in polynomial time. Then the problem $\mathcal{P}_2$ with input N has a "yes" answer if and only if the problem $\mathcal{P}_1$ with input $aX^2 + bX + c$ has a "yes" answer.

Definition 4.3 can also be used in a converse way. Suppose that we know (or believe) the problem $\mathcal{P}_1$ to be very difficult. That is, we are virtually certain that there is no efficient algorithm for it. If $\mathcal{P}_1$ reduces to $\mathcal{P}_2$, then it follows that there is no efficient algorithm for $\mathcal{P}_2$ either.

Definition 4.3 is a little too restrictive. It is worthwhile to have a broader definition of polynomial time reduction of $\mathcal{P}_1$ to $\mathcal{P}_2$ that allows us to use several different instances of P_2 to solve a single instance of P_1. To give such a definition, we first say what is meant by an "oracle".

Definition 4.4. Let P_2 be a decision or search problem. In describing an algorithm for some other problem P_1, whenever we say "call to a P_2-oracle" we mean that our algorithm has created an instance of P_2, and we suppose that some other algorithm then gives us the corresponding P_2-output. The time taken by this algorithm for P_2 is not included in the running time for the algorithm for P_1. In other words, we pretend that the algorithm for P_2 is a "black box" that works instantaneously.

To use the language of computer programming, we can think of an oracle as a subroutine that we are free to call upon without including its running time in our estimate for the total running time of our program.

Definition 4.5. Let P_1 and P_2 be two problems (either decision problems or search problems). We say that P_1 *reduces to* P_2 *in polynomial time* if there is a polynomial time algorithm for P_1 that uses at most polynomially many calls to a P_2-oracle.

Example 4.11. Let P_1 be the Integer Factorization search problem, and let P_2 be the Integer Factorization decision problem. We saw that P_1 can be solved with n calls to a P_2-oracle, where n is the input length. (Example 4.7 goes through the procedure in the case when $N = 91$, $n = 7$.) Thus, P_1 reduces to P_2.

4.4 NP-Complete Problems

Definition 4.6. A decision problem P in NP is said to be "NP-complete" if every other problem Q in NP can be reduced to P in polynomial time.

To put it another way, if one had a polynomial time algorithm for an NP-complete problem P, then one would also have polynomial time algorithms for all other NP problems Q. This would mean that P equals NP, and the P$\neq$NP conjecture would be false. For this reason, no one is likely to come up with a polynomial time algorithm for any NP-complete problem. In a sense, the NP-complete problems are the most difficult problems in the class NP.

This statement should be taken cautiously. One might be able to produce an efficient algorithm – even a polynomial time algorithm – that gives an answer to *most* instances of a certain NP-complete problem. This would not contradict the P$\neq$NP conjecture.

In practice, though, it is usually hard to efficiently solve large instances of an NP-complete problem. Often the running times of all known algorithms grow exponentially with the length of the input.

It can be shown that the Traveling Salesrep problem in Example 4.5 is NP-complete (see [Garey and Johnson 1979]). Suppose that one has a very complicated instance with several hundred cities – say, the business class airfare map for the U.S. It might take millions of millions of years – longer than the lifetime of the

Universe – for the fastest computers to find an optimal solution to this instance of the Traveling Salesrep problem.

It can also be shown that 3-Coloring is NP-complete.

Finally, note that it is possible for a problem $\mathcal{P}$ to reduce to an NP-problem even though $\mathcal{P}$ itself is not likely to be in NP.

Example 4.12. The Exact Traveling Salesrep problem is the following decision problem.

INPUT: A set of cities and distances between them, and an integer k.

QUESTION: Does a shortest tour through all the cities have length exactly equal to k?

To provide a certificate for a "yes" answer it would not be enough simply to show a tour of length k; the extraterrestrial in Definition 4.2 would have to somehow convince us that there is no shorter tour. It is very unlikely that this could be done. (The existence of such a certificate would mean that NP=co-NP; see the answer to Exercise 11(b) below.) On the other hand, one can use the binary search method to reduce the Exact Traveling Salesrep decision problem to the Traveling Salesrep decision problem. It is also possible (but a little harder – see [Papadimitriou 1994], p. 411–412) to show the converse: that Traveling Salesrep reduces to Exact Traveling Salesrep. In such a case we say that the two problems are *polynomial time equivalent* (sometimes the term "NP-equivalent" is used).

We conclude this section with a definition that applies to problems that are not necessarily in NP.

Definition 4.7. A decision or search problem is said to be *NP-hard* if any NP-problem reduces to it.

Because of the transitivity of reduction, to show that a problem is NP-hard it suffices to find a single NP-complete problem that reduces to it.

Exercises for §4

1. Arrange these problems from lowest to highest according to input length:
(a) the problem of multiplying together 20 integers, each $\approx 10^{100}$;
(b) a Traveling Salesrep problem with 20 cities, where all of the distances are integers between 1 and 100;
(c) the problem of finding the roots of a quadratic polynomial whose coefficients are integers of about 50 digits;
(d) the problem of finding all prime factors of an integer of about 40 digits.

2. For which of the problems (a)–(d) in Exercise 1 do you know a polynomial time algorithm?

3. Using the big-O notation, find an upper bound in terms of B for the input length of the Traveling Salesrep problem if the number of cities is at most B and the distance between any two cities is also at most B.

4. Explain why the 3-Coloring problem for maps may be regarded as a special case of the 3-Coloring problem for graphs.

5. Explain how to use an algorithm for the Traveling Salesrep decision problem to solve the Traveling Salesrep search problem.

6. Suppose that P_1 is the problem

> INPUT: Two integers.
> QUESTION: Are they equal?

Suppose that P_2 is the problem

> INPUT: Two equations $ax+by = 0$ and $cx+dy = 0$, where a, b, c, d are integers.
> QUESTION: Do these equations have any common solutions (x, y) other than $(0, 0)$?

Show that P_2 reduces to P_1 by constructing a reduction of instances of one problem to instances of the other.

7. Suppose that P_1 is the problem

> INPUT: Two vectors in 3-dimensional space.
> QUESTION: Are they proportional?

Suppose that P_2 is the problem

> INPUT: Two pairs of (non-proportional) vectors in 3-dimensional space.
> QUESTION: Do both pairs of vectors span the same plane?

Show that P_2 reduces to P_1 by constructing a reduction of instances of one problem to instances of the other.

8. Let P_1 be the problem

> INPUT: A polynomial $p(X)$ with integer coefficients.
> QUESTION: Is there any interval of the real number line on which $p(X)$ decreases?

Let P_2 be the problem

> INPUT: A polynomial $p(X)$ with integer coefficients.
> QUESTION: Is there any interval of the real number line on which $p(X)$ is negative?

Show that P_1 reduces to P_2.

9. Let P_1 be the following search problem:

> INPUT: Two integers e and N, where $N > 1$ is odd.
> OUTPUT: An integer d such that the map $x \mapsto x^d$ modulo N inverts the map $x \mapsto x^e$ modulo N for all integers x prime to N, provided that such d exists; if no such d exists, then the statement that such a d cannot be found.

Let P_2 be:

INPUT: An odd integer $N > 1$.
OUTPUT: A nontrivial factor M of N, or else the statement that N is prime.

Show that P_1 reduces to P_2 in the sense of Definition 4.5. P_2 is the Integer Factorization search problem, and P_1 is the RSA problem. It is not known whether P_2 reduces to P_1 (in which case the two problems would be "polynomial time equivalent").

10. Let p be a fixed prime, and let g be a fixed integer not divisible by p. Let P_1 be the following search problem:

INPUT: Two integers a and b.
OUTPUT: (1) If there exist integers k and l such that $a \equiv g^k$ (modulo p) and $b \equiv g^l$ (modulo p), then give the least positive residue of g^{kl} modulo p. (2) If no such k and l exist, then state that a and/or b is not a power of g modulo p.

Let P_2 be the following search problem:

INPUT: An integer a.
OUTPUT: An integer k such that $a \equiv g^k$ (modulo p), if such k exists; otherwise, the statement that no such k exists.

Show that P_1 reduces to P_2 in the sense of Definition 4.5. P_1 is called the Diffie–Hellman problem, and P_2 is called the Discrete Logarithm problem. It is not known whether P_2 reduces to P_1, that is, whether the two problems are polynomial time equivalent. In recent years important partial results have been proved that support the conjecture that P_1 is equivalent to P_2. See, for example, [Boneh and Lipton 1996].

11. Are the following decision problems likely to be in NP? Explain.

(a) INPUT: A positive integer N.
QUESTION: Is $\pi(N)$ an even number?

Recall that $\pi(N)$ denotes the number of primes less than or equal to N.

(b) INPUT: A list of cities and distances between any two cities, and an integer k.
QUESTION: Do all tours that pass through all of the cities have length greater than k?

(c) INPUT: A graph and an integer k.
QUESTION: Does the graph have k or more different 3-colorings?

12. If a subexponential time algorithm (see Definition 3.2) is found for an NP-complete problem, then any NP problem has a subexponential time algorithm. True or false? Explain.

§5. Promise Problems

5.1 The Cracking Problem

Suppose that we are trying to cryptanalyze a public key cryptosystem. That is, we know the public enciphering key E and the one-to-one function f_E from the set $\mathcal{P}$ of plaintext message units to the set $\mathcal{C}$ of ciphertext message units. We intercept some $y \in \mathcal{C}$, and we want to determine the unique $x \in \mathcal{P}$ such that $f_E(x) = y$. This is known as the *cracking problem* for a public key cryptosystem. That is, the cracking problem is as follows:

INPUT: E, $f_E : \mathcal{P} \longrightarrow \mathcal{C}$, $y \in \mathcal{C}$.
OUTPUT: $x \in \mathcal{P}$ such that $f_E(x) = y$.

Unlike the problems in the last section, the cryptanalyst knows something other than the input. Namely, she knows that there exists $x \in \mathcal{P}$ such that $f_E(x) = y$ (in other words, y is contained in the image of the function), and, moreover, x is unique. Thus, the cracking problem is of a slightly different sort from our earlier examples, and so one needs a new definition that captures this situation.

Definition 5.1. A *promise* problem is a search or decision problem with a condition attached to it. When analyzing an algorithm for a promise problem, we disregard what happens when the condition fails – that is, we do not care if the algorithm gives the wrong answer or no answer at all in such a case.

Example 5.1. It is natural to regard the cracking problem for any public key encryption system as a promise problem. As before, we suppose that we know the enciphering key E and the function $f_E : \mathcal{P} \longrightarrow \mathcal{C}$.

INPUT: E, $f_E : \mathcal{P} \longrightarrow \mathcal{C}$, $y \in \mathcal{C}$.
PROMISE: f_E is one-to-one, and y is in the image of f_E.
OUTPUT: $x \in \mathcal{P}$ such that $f_E(x) = y$.

Example 5.2. The following is a promise version of the Integer Factorization search problem:

INPUT: An integer $N > 1$.
PROMISE: N is a product of two distinct prime numbers.
OUTPUT: A nontrivial factor M of N.

An efficient algorithm for this promise problem would suffice to break RSA. Of course, in the unlikely event that an efficient algorithm was found that could factor a product of two primes but not a product of three primes, it would be easy to modify RSA to use moduli that are products of three primes.

5.2 The Trapdoor Problem

Example 5.2 is of a different sort from Example 5.1, although it is also related to the problem of cryptanalyzing a public key encryption system. Example 5.2 is concerned with solving the underlying mathematical problem upon which RSA is based, whereas the cracking problem is the narrowest possible description of the cryptanalyst's task.

It is important to distinguish between these two types of problems. We shall use the term *trapdoor problem* for a promise problem that asks us to reverse the basic mathematical construction of the trapdoor in a public key cryptosystem. If we find an efficient algorithm for the trapdoor problem, then we will also have an efficient algorithm for the cracking problem.

The converse is not necessarily true (and even if true, it might be very hard to prove). That is, there might be ways of solving the cracking problem without dealing directly with the underlying trapdoor function. No one has been able to prove, for instance, that the only way to break RSA is to factor the modulus.

§ 6. Randomized Algorithms and Complexity Classes

6.1 An Example

Example 6.1. The randomized algorithm that follows is one of the best ways of testing an odd number N to see whether it is prime. More precisely, the test will determine either that (1) N is *probably prime*, or else (2) N is *definitely* composite.

First we write $N - 1$ in the form $N - 1 = 2^s t$, where 2^s is the largest power of 2 dividing $N - 1$ and t is an odd number. We randomly choose a number a with $1 < a < N - 1$. Then we raise a to the $(N - 1)$-st power modulo N in two stages: (i) we find the least nonnegative residue of a^t modulo N by the "square and multiply" method (see Example 3.5); and (ii) we successively square a^t modulo N until we get $a^{2^s t} = a^{N-1}$:

$$a^t \bmod N, \quad a^{2t} \bmod N, \quad a^{4t} \bmod N, \ldots,$$
$$\ldots, a^{2^{s-1}t} \bmod N, \quad a^{N-1} \bmod N . \tag{1}$$

Now we note that if N is a prime number, then

1) $a^{N-1} \equiv 1 \pmod{N}$ (this is Fermat's Little Theorem), so that the last number in (1) will be 1;
2) if not all the numbers in (1) are 1, then the first 1 in the list will be preceded by $N - 1$ (this is because the only square roots of 1 modulo a prime number N are ± 1).

When both 1) and 2) hold, we say that N *passes the strong Fermat primality test to the base* a (also called "Miller's test" and "Rabin's probabilistic primality test"). If N fails either 1) or 2), then it is definitely composite. If, however, N

passes the strong primality test to the base a, then there is "a greater than 75% chance that N is prime". By this we mean that if N is composite, then it passes the strong primality test to the base a for fewer than 25% of all a in the range $1 < a < N - 1$ (for a proof of this fact, see [Rosen 1993], p. 302–305).

If the strong Fermat primality test is performed for k different randomly chosen values of a, and if N satisfies 1) and 2) for all of these a, then we can say that there is "at least a $1 - 4^{-k}$ probability that N is prime".

6.2 The Complexity Class RP

This complexity class captures what is going on in Example 6.1.

Definition 6.1. We say that a decision problem $\mathcal{P}$ is *solvable in randomized polynomial time* (or "probabilistic polynomial time") and we write $\mathcal{P} \in$ RP if there exists a polynomial time algorithm that includes a random selection of one or more integers and, depending on that random choice, produces a "yes" or "no" answer, where in the former case the "yes" answer is definitely correct and in the latter case there is a probability greater than 1/2 that the "no" answer is correct.

In Example 6.1 we saw that the following Compositeness problem is in RP:
INPUT: A positive odd integer N.
QUESTION: Is N a composite number?

Example 6.2. Another example of a problem in RP is Product Polynomial Inequivalence.

INPUT: Two sets of polynomials $\{P_1, \ldots, P_m\}$ and $\{Q_1, \ldots, Q_n\}$, where each of the P_i and Q_j is a polynomial in one or several variables with rational coefficients.
QUESTION: Are $\prod_{i=1}^{m} P_i$ and $\prod_{j=1}^{n} Q_j$ different polynomials?

Here each polynomial in the input is listed by giving its terms with nonzero coefficients. If the polynomials are "sparse" – that is, if most of their coefficients are zero – then the input length will be much less than if the polynomials were given by listing all of the terms (including the ones with zero coefficient) in lexicographical order.

Notice that the running time for the obvious method of answering the question – by simply multiplying out both sets of polynomials – is not generally bounded by a polynomial in the input length. In fact, the number of nonzero terms in each product polynomial might be exponentially large as a function of the input length.

However, there is a simple method to test whether or not $\prod P_i = \prod Q_j$. Suppose that the P_i and Q_j are polynomials in l variables $X_1, \ldots, X_l$. In some random way choose l rational numbers $x_1, \ldots, x_l$, and evaluate each of the polynomials at $X_k = x_k$, $k = 1, \ldots, l$. Then determine whether or not

$$\prod_{i=1}^{m} P_i(x_1, \ldots, x_l) = \prod_{j=1}^{n} Q_j(x_1, \ldots, x_l) \ .$$

If not, then you know that the two products of polynomials are unequal; that is, the answer to Product Polynomial Inequivalence is *definitely* "yes". If, on the other hand, the above products of rational numbers are equal, then the answer is *probably* "no". Of course, one cannot be sure that two polynomials are identically equal just because their values at a particular point are equal. But if their values are equal at a large number of randomly chosen points, then one can say that they are almost certain to be equal – that there is a probability at least $1 - \varepsilon$ that "no" is the correct answer (where ε is a constant that does not depend on the input). Thus, Product Polynomial Inequivalence is in the complexity class RP.

Remark. If $\mathcal{P} \in$ RP, then for any constant $\varepsilon > 0$ one has an algorithm whose "no" answers have a probability greater than $1 - \varepsilon$ of being correct. It suffices to take k independent iterations of the algorithm in the definition, where k is chosen so that $2^{-k} < \varepsilon$.

6.3 The Complexity Class BPP

Definition 6.2. We say that a decision problem $\mathcal{P}$ is *solvable with probability bounded away from* $1/2$ *in randomized polynomial time* and we write $\mathcal{P} \in$ BPP if there exist a constant $\delta > 0$ and a polynomial time algorithm that includes a random selection of one or more integers and, depending on that random choice, produces a "yes" or "no" answer, where in either case the probability that the answer is correct is greater than $1/2 + \delta$.

Remark. Just as in the case of RP (see the previous remark), if $\mathcal{P} \in$ BPP, then for any constant $\varepsilon > 0$ one has an algorithm whose answers have a probability greater than $1 - \varepsilon$ of being correct. Namely, we consider a new algorithm consisting of k iterations of the algorithm in the definition, followed by a "vote": the answer to the new algorithm is "yes" if and only if the answer to more than $k/2$ of the iterations was "yes". Using standard techniques of probabilities and statistics, one can show that for any constant δ there exists a constant k such that there is a probability greater than $1 - \varepsilon$ that the "vote" algorithm gives the correct answer. This is intuitively obvious if we think of a weighted coin that has $1/2 + \delta$ probability of landing "heads" rather than "tails". If we toss the coin a sufficiently large number of times, there is a greater than 99.9% chance that heads will come up more than tails.

The definition of the class BPP is fairly broad. For instance, BPP contains RP (see Exercise 3 below) and is probably much larger. Yet Definition 4.2 is still stringent enough to guarantee* that we have a practical algorithm for the problem.

---

* The word "guarantee" is too strong here. See the discussion of P and practicality following Definition 4.1.

Exercises for § 6

1. As a function of the input length in the Compositeness problem, what is the order of magnitude of time required for a single strong Fermat primality test?

2. Give a simple example of a set of polynomials $\{P_1, \ldots, P_m\}$ such that the number of terms in $\prod_{i=1}^m P_i$ (after the product is multiplied out) is exponentially large compared to the total number of nonzero terms in $P_1, \ldots, P_m$.

3. Explain why BPP⊃RP∪co-RP. Here co-RP denotes the set of decision problems that satisfy the definition of RP with "yes" and "no" reversed. For example, the following Primality problem, which is the reverse of the Compositeness problem, is in co-RP:

INPUT: A positive odd integer N.
QUESTION: Is N a prime number?

4. Explain the difference between the sense in which you can solve a problem $\mathcal{P}$ if $\mathcal{P} \in$ RP ∪ co-RP and the sense in which you can solve $\mathcal{P}$ if $\mathcal{P} \in$ RP ∩ co-RP. (The latter class is often denoted ZPP. In [Adleman and Huang 1992] it is shown that Primality belongs to ZPP.)

§ 7. Some Other Complexity Classes

7.1 The Polynomial Hierarchy

Let $\Delta_1 = $ P and let $\Sigma_1 = $ NP.

Definition 7.1. Define Δ_2 to be P^{NP}, which is the class of decision problems that have a polynomial time reduction (in the sense of Definition 4.5) to a problem in NP. In other words, $\mathcal{P} \in \Delta_2$ if there is a problem $\mathcal{Q} \in$ NP and a polynomial time algorithm for $\mathcal{P}$ that makes at most polynomially many calls on a $\mathcal{Q}$-oracle.

Definition 7.2. Define Σ_2 to be NP^{NP}. This is the class of decision problems whose "yes" answers have a certificate that can be verified by a polynomial time algorithm using an oracle for a problem in NP. That is, an extraterrestrial with unlimited computing power could, in the case of a "yes" answer to an instance of the problem, produce a certificate such that the following holds: there is a problem $\mathcal{Q} \in$ NP and a polynomial time algorithm with at most polynomially many calls on a $\mathcal{Q}$-oracle that you can use to verify the certificate and thereby be certain that the "yes" answer is correct.

Notice the difference between NP and Σ_2. In the former we need certificates that can be checked in absolute polynomial time. In the latter we are allowed to use an oracle for any fixed NP-problem in our algorithm for checking the certificate.

The classes Δ_2 and Σ_2 make up the "second level of the polynomial hierarchy", along with Π_2, which is the co-class of Σ_2, i.e., the class of problems that belong to Σ_2 after the question is reversed (so that the answer is "yes" whenever the answer to the original problem is "no", and vice-versa). The successive levels of the hierarchy are defined inductively as follows: $\Delta_{k+1} = \mathrm{P}^{\Sigma_k}$, $\Sigma_{k+1} = \mathrm{NP}^{\Sigma_k}$, and $\Pi_{k+1} = \mathrm{co}\text{-}\Sigma_{k+1}$, for $k = 2, 3, \ldots$. In other words, each level is constructed using oracles for problems from the previous level.

Definition 7.3. The union of all of these classes, denoted PH, is called the "polynomial hierarchy".

It is easy to see that each level of PH is contained in the next level. It has not been proved that any of these containments are proper; it is conjectured that they all are. If the P≠NP conjecture turns out to be false, then the entire polynomial hierarchy "collapses"; in other words, PH=P in that case. The study of PH itself is mainly of theoretical interest; from a practical point of view, there are very few interesting problems in PH that lie above the second level of the hierarchy.

7.2 Unique P

Definition 7.4. The class UP ("unique P") consists of NP-problems for which there exists a prescription for a uniquely determined polynomial time certificate for any instance having a "yes" answer.

For example, the Traveling Salesrep decision problem is *not* likely to belong to UP. The obvious certificate for a "yes" answer – a description of a tour of length $\leq k$ – is not, in general, unique. (See the remark preceding Example 4.4.)

On the other hand, any one-to-one function $f : X \longrightarrow Y$ that can be computed in polynomial time gives a corresponding problem in UP, namely:

INPUT: $y \in Y$.
QUESTION: Is there an $x \in X$ such that $f(x) = y$?

The certificate for a "yes" answer consists simply of the unique x for which $f(x) = y$. It can be shown that the one-way encryption functions of public key cryptography (see Definition 2.1 of Chapter 1) exist if and only if UP is strictly larger than P.

It is obvious that P⊂UP⊂NP; however, neither of these inclusions has been proven to be a strict inclusion. Of course, a proof either that P≠UP or that UP≠NP would also be a proof of the fundamental P≠NP conjecture.

7.3 Average Time

The notions of complexity discussed thus far all relate to the *worst case* of a problem. For example, an NP-problem $\mathcal{P}$ is NP-complete if, roughly speaking, an algorithm that efficiently solves all instances of $\mathcal{P}$ – including the most difficult ones – would lead to an efficient algorithm for any other NP-problem $\mathcal{Q}$. But suppose that we have an algorithm that efficiently solves *most* instances of an NP-problem $\mathcal{P}$ (with some reasonable definition of the word "most"). That might be enough for our practical applications. However, this would not necessarily imply that we have a useful algorithm for Q. It might turn out that our method of reducing Q to P usually leads to instances of P that are not included in the "most" – that is, the instances of Q that are of practical interest might reduce to instances of P that the algorithm cannot solve efficiently.

In cryptographic applications it is not really enough to know that the hardest instances of the cracking problem or the trapdoor problem (see §§5.1–5.2) are hard. What one wants to know is that "most" instances (or most instances constructed so that some additional conditions hold) are hard. For example, the security of RSA is based on the assumption that most numbers obtained as the product of two randomly chosen large primes are hard to factor.

How could one give a precise definition that captures this notion? The following definition is due to Levin [1984].

Definition 7.5. Let P be a decision or search problem, and let μ_n be a distribution on the set of all instances of input length at most n. That means that μ_n is a function that assigns a non-negative real number to any instance having input length at most n, and the sum of the values of μ_n on all such instances is equal to 1. We say that P is *polynomial time on average* with respect to the distributions μ_n if we have an algorithm that solves P and has the following property: for some $\varepsilon > 0$

$$\sum_i T(i)^\varepsilon \mu_n(i) = O(n) \qquad \text{as } n \to \infty ,$$

where $T(i)$ is the time the algorithm takes to solve the instance i, and the sum is taken over all instances i of input length n or less.

Intuitively, if S is a set of instances of P of input length at most n, then $\left(\sum_{i \in S} \mu_n(i)\right)^{-1}$ can be thought of as a measure of the amount of time it takes to find an instance in S by random sampling. Roughly speaking, Definition 7.5 says that for some $\varepsilon > 0$ it will take time at least $f(n)^\varepsilon$ to find an instance of input length n for which the running time of the algorithm is $\geq f(n)$, where $f(n)$ is a rapidly growing function of n, such as n^k for k large. (See Exercise 2 below.)

For more discussion of Levin's definition and other aspects of average-case complexity and cryptography, see [Ben-David, Chor, Goldreich, and Luby 1989] and [Impagliazzo 1995].

7.4 Interaction

Both cryptographic protocols and games are characterized by *interaction* between two or more "players". Various complexity classes have been defined to deal with such situations. One of the most important is the class IP of problems solvable by *interactive proof systems*.

Definition 7.6. The class IP consists of all decision problems that are solvable by a procedure involving two players, one of whom (the "prover") has unlimited computational power and the other of whom (the "verifier") has a source of random bits and is subject to a polynomial bound on total computation time.

In an interactive proof system or the closely related Arthur–Merlin protocol,[*] the verifier Arthur and the prover Merlin exchange messages. Merlin is like the advanced extraterrestrial being in the definition of NP. He is presumed to have unlimited computing power. He must convince Arthur that a certain instance of a decision problem has answer 'yes.' Unless the problem is in NP, Merlin won't be able to convince Arthur by simply sending him a single message (a polynomial time certificate of a 'yes' answer – see Definition 4.2). Rather, it will probably take a fairly lengthy interchange of messages before Arthur is convinced beyond a reasonable doubt that the answer is 'yes.' Arthur might toss a coin and make queries of Merlin. If he is satisfied with Merlin's responses, eventually he will have to admit that there is a probability less than ε that Merlin could have given those answers if the correct answer were 'no.'

An important and unexpected result of Shamir [1992] is that the class IP in Definition 7.6 is the same as the class PSPACE, which consists of all decision problems that can be solved by an algorithm with a polynomial bound on the allowed memory but no bound on time. PSPACE is very large: it includes all of PH (see Definition 7.3) and is probably much bigger.

There are complexity classes that are much, much larger even than the class IP=PSPACE. For example, EXPSPACE is the class of all decision problems that can be solved by an algorithm with an exponential bound on the allowed memory. We shall later (see §4 of Chapter 5) encounter a problem $\mathcal{P}$ that is "EXPSPACE-hard". This term means that any problem in EXPSPACE can be reduced to $\mathcal{P}$ in polynomial time.

7.5 Parallelism, Non-uniformity

The complexity concepts in the earlier sections do not cover all possible features of an algorithm one might construct to break a cryptosystem. For example, some problems can be solved using massively parallel computations. That is, algorithms are known that can be greatly speeded up if we have a vast number of computers all working at once. In other cases, all of the known algorithms have to be carried out largely *in series* rather than *in parallel*, and so it would not help much to have several processors simultaneously working on the problem.

We will not dwell on massively parallel complexity classes, because thus far they have not been of great importance in cryptography. However, we shall give one definition in order to give the flavor of the subject.

Definition 7.7. The class NC consists of decision problems for which there exist constants C_1 and C_2 and a deterministic algorithm that can solve an instance with input length n in time bounded by $\ln^{C_1} n$ using at most n^{C_2} processors at the same time.

Another concept that has relevance to cryptography is *non-uniformity*.

---

[*] The name comes from the legendary King Arthur and his wizard Merlin.

Definition 7.8. We say that a problem $\mathcal{P}$ is solvable in *non-uniform polynomial time* and we write $\mathcal{P} \in$ P/poly if there exists a polynomial $p(n)$ and a sequence of algorithms A_n such that the algorithm A_n will solve all instances of input length at most n in time bounded by $p(n)$.

To show that a problem is in the class P/poly we must give a recipe for the algorithms A_n. Notice that Definition 7.8 says nothing about the length of time this recipe takes. For example, the set-up of these algorithms might require a lengthy "pre-computation" whose running time grows exponentially in n. However, once A_n is set up, it will be able to quickly solve an arbitrary instance of input length $\leq n$.

In practice, it often happens that we know in advance that we will want to solve instances of a problem $\mathcal{P}$ whose input lengths vary in a small range, for example, $100 \leq n \leq 150$. We might be willing to go to tremendous effort to set up an algorithm that works only for $n \leq 150$. The time and money to do this must be spent just once, after which the algorithm handles our needs cheaply and efficiently. To some extent Definition 7.8 captures this situation.

Exercises for § 7

1. Prove that the trial division algorithm for factorization is not polynomial time on average. More precisely, consider the problem

   INPUT: A positive odd integer N.
   OUTPUT: The smallest prime divisor of N.

Define the distribution μ_n on all instances of binary length $\leq n$ as follows: $\mu_n(i) = 1/2^{n-1}$. In the trial division algorithm one divides all odd numbers $3, 5, 7, \ldots$ into N until one either finds a divisor of N or else reaches $\sqrt{N}$ (in the latter case N is prime, and the output is N).

2. Suppose that for all $\varepsilon > 0$ there exists $k > 1/\varepsilon$ such that for all $n \geq n_0$ by random sampling in time $n^{k\varepsilon}$ we can find an instance of $\mathcal{P}$ of input length $\leq n$ for which our algorithm takes time greater than n^k. Show that $\mathcal{P}$ is not polynomial time on average with respect to this algorithm in the sense of Definition 7.5.

3. Suppose that Levin's property in Definition 7.5 were replaced by the following slightly simpler statement: $\sum T(i)\mu_n(i) = O(n^c)$ for some constant c. Show that a function $T(i)$ that satisfies this property also satisfies Levin's property, but the converse is false.

4. Show that the following problem is in NC. The input consists of two polynomials with integer coefficients, where the maximum absolute value of the coefficients is less than the degree. The input is given by listing all of the coefficients (not only the nonzero ones). The output is the sum of the two polynomials.

5. Show that NC$\subset$P (see Definition 7.8).

Chapter 3. Algebra

§ 1. Fields

We start out by recalling the basic definitions and properties of a field.

Definition 1.1. A *field* is a set $\mathbb{F}$ with *multiplication* and *addition* operations that satisfy the familiar rules – associativity and commutativity of both addition and multiplication, the distributive law, existence of an additive identity 0 and a multiplicative identity 1, additive inverses, and multiplicative inverses for everything except 0.

The following fields are basic in many areas of mathematics: (1) the field $\mathbb{Q}$ consisting of all rational numbers; (2) the field $\mathbb{R}$ of real numbers; (3) the field $\mathbb{C}$ of complex numbers; (4) the field $\mathbb{Z}/p\mathbb{Z}$ of integers modulo a prime number p. The latter field is often denoted $\mathbb{F}_p$, and in some places it is denoted $GF(p)$.

Definition 1.2. A *vector space* can be defined over any field $\mathbb{F}$ by the same properties that are used to define a vector space over the real numbers. Any vector space has a *basis*, and the number of elements in a basis is called its *dimension*. An *extension field*, by which we mean a bigger field containing $\mathbb{F}$, is automatically a vector space over $\mathbb{F}$. We call it a *finite extension* if it is a finite dimensional vector space. By the *degree* of a finite extension we mean its dimension as a vector space. One common way of obtaining extension fields is to *adjoin* an element to $\mathbb{F}$: we say that $\mathbb{K} = \mathbb{F}(\alpha)$ if $\mathbb{K}$ is the field consisting of all rational expressions formed using α and elements of $\mathbb{F}$.

Definition 1.3. The *polynomial ring* over the field $\mathbb{F}$ in the set of variables $X = \{X_1, \ldots, X_m\}$, denoted $\mathbb{F}[X]$, consists of all finite sums of products of powers of $X_1, \ldots, X_m$ with coefficients in $\mathbb{F}$. (When $m = 2$ we often use X and Y instead of X_1 and X_2; and if $m = 3$ we often use X, Y, Z.) One adds and multiplies polynomials in $\mathbb{F}[X]$ in the same way as one does with polynomials over the reals. We say that g *divides* f, where f, $g \in \mathbb{F}[X]$, if there exists a polynomial $h \in \mathbb{F}[X]$ such that $f = gh$. The *irreducible* polynomials $f \in \mathbb{F}[X]$ are those such that the relation $f = gh$ implies that either g or h is a constant; they play the role among the polynomials that the prime numbers play among the integers. In this section all polynomials will be in one variable, so we take $m = 1$ and $X = X_1$. The *degree* d of a polynomial in one variable is the highest power of X

that occurs with nonzero coefficient. We say that the polynomial is *monic* if the coefficient of X^d is 1.

Polynomial rings (in one or more variables) have *unique factorization*, meaning that every polynomial in $\mathbb{F}[X]$ can be written in one and only one way (except for constant terms and the order of factors) as a product of irreducible elements of $\mathbb{F}[X]$.

Definition 1.4. An element α in some extension field $\mathbb{K}$ containing $\mathbb{F}$ is said to be *algebraic* over $\mathbb{F}$ if there is a polynomial in one variable $f(X) \in \mathbb{F}[X]$ such that $f(\alpha) = 0$. In that case there is a *unique* monic irreducible polynomial in $\mathbb{F}[X]$ of which α is a root (and any other polynomial that α satisfies must be divisible by this monic irreducible polynomial). This monic irreducible polynomial is called the *minimal polynomial of α*.

If the minimal polynomial of α has degree d, then any element of $\mathbb{F}(\alpha)$ (that is, any rational expression involving powers of α and elements of $\mathbb{F}$) can be expressed as a linear combination of the powers 1, α, $\alpha^2, \ldots, \alpha^{d-1}$. Thus, those powers of α form a basis of $\mathbb{F}(\alpha)$ over $\mathbb{F}$, and so the degree of the extension obtained by adjoining α is the same as the degree of the minimal polynomial of α.

Definition 1.5. Any other root α' of the minimal polynomial of α is called a *conjugate* of α over $\mathbb{F}$. The product of all of the conjugates of α (including α itself) is called its *norm*.* If α' is a conjugate of α, then the fields $\mathbb{F}(\alpha)$ and $\mathbb{F}(\alpha')$ are *isomorphic* by means of the map that takes any expression in terms of α to the same expression with α replaced by α'. The word "isomorphic" means that we have a 1-to-1 correspondence between the two fields that preserves addition and multiplication. If it happens that $\mathbb{F}(\alpha)$ and $\mathbb{F}(\alpha')$ are the same field, we say that the map that takes α to α' gives an *automorphism* of the field.

For example, $\sqrt{2}$ has one conjugate over $\mathbb{Q}$, namely $-\sqrt{2}$, and the map $a + b\sqrt{2} \mapsto a - b\sqrt{2}$ is an automorphism of the field $\mathbb{Q}(\sqrt{2})$ (which consists of all real numbers of the form $a + b\sqrt{2}$ with a and b rational).

Definition 1.6. The *derivative* of a polynomial in one variable and the *partial derivatives* of a polynomial in several variables are defined using the nX^{n-1} rule (not as a limit, since limits don't make sense unless there is a concept of distance or a topology in $\mathbb{F}$).

A polynomial f of degree d in one variable X may or may not have a root $r \in \mathbb{F}$, that is, such that $f(r) = 0$. If it does, then the degree-1 polynomial $X - r$ divides f; if $(X - r)^m$ is the highest power of $X - r$ that divides f, then we say that r is a root of *multiplicity* m. Because of unique factorization, the total number of roots of f in $\mathbb{F}$ counting multiplicity cannot exceed d. If a polynomial

---

* Here we are assuming that our field extension is *separable*, as is always the case in this book. Algebraic extensions of finite fields and extensions of fields of characteristic zero are always separable.

$f \in \mathbb{F}[X]$ has a multiple root r, then r will be a root of both f and its derivative f', and hence a root of the *greatest common divisor* (see §3) of f and f', which is denoted g.c.d.(f, f').

Definition 1.7. Given any polynomial $f(X) \in \mathbb{F}[X]$ in one variable, there is an extension field $\mathbb{K}$ of $\mathbb{F}$ such that $f(X) \in \mathbb{K}[X]$ splits into a product of linear factors (equivalently, has d roots in $\mathbb{K}$ counting multiplicity, where d is its degree) and such that $\mathbb{K}$ is the smallest extension field containing those roots. $\mathbb{K}$ is called the *splitting field* of f. The splitting field is unique *up to isomorphism*, meaning that if we have any other field $\mathbb{K}'$ with the same properties, then there must be a 1-to-1 correspondence $\mathbb{K} \xrightarrow{\sim} \mathbb{K}'$ that preserves addition and multiplication.

For example, $\mathbb{Q}(\sqrt{2})$ is the splitting field of $f(X) = X^2 - 2 \in \mathbb{Q}[X]$. To obtain the splitting field of $f(X) = X^3 - 2 \in \mathbb{Q}[X]$ one must adjoin to $\mathbb{Q}$ both $\sqrt[3]{2}$ and $\sqrt{-3}$. (Recall that the nontrivial cube roots of 1 are $(-1 \pm \sqrt{-3})/2$, so that adjoining $\sqrt{-3}$ is equivalent to adjoining all cube roots of 1.)

Definition 1.8. If a field $\mathbb{F}$ has the property that every polynomial with coefficients in $\mathbb{F}$ factors completely into linear factors, then we say that $\mathbb{F}$ is *algebraically closed*. Equivalently, it suffices to require that every polynomial with coefficients in $\mathbb{F}$ have a root in $\mathbb{F}$. For instance, the field $\mathbb{C}$ of complex numbers is algebraically closed.

The smallest algebraically closed extension field of $\mathbb{F}$ is called the *algebraic closure* of $\mathbb{F}$. It is denoted $\overline{\mathbb{F}}$. For example, the algebraic closure of the field of real numbers is the field of complex numbers.

Definition 1.9. If adding the multiplicative identity 1 to itself in $\mathbb{F}$ never gives 0, then we say that $\mathbb{F}$ has *characteristic zero*; in that case $\mathbb{F}$ contains a copy of the field of rational numbers. Otherwise, there is a prime number p such that $1 + 1 + \cdots + 1$ (p times) equals 0, and p is called the *characteristic* of the field $\mathbb{F}$. In that case $\mathbb{F}$ contains a copy of the field $\mathbb{Z}/p\mathbb{Z}$, which is called its *prime field*.

Exercises for §1

1. Let $\mathbb{K}$ be the splitting field of the polynomial $X^3 - 2$ over $\mathbb{F}$. Find the degree of $\mathbb{K}$ if $\mathbb{F}$ is (a) $\mathbb{Q}$; (b) $\mathbb{R}$; (c) $\mathbb{F}_5 = \mathbb{Z}/5\mathbb{Z}$; (d) $\mathbb{F}_7 = \mathbb{Z}/7\mathbb{Z}$; (e) $\mathbb{F}_{31} = \mathbb{Z}/31\mathbb{Z}$. Explain your answers.

2. Prove that a polynomial in $\mathbb{F}_p[X]$ has derivative identically zero if and only if it is the p-th power of a polynomial in $\mathbb{F}_p[X]$. Give a criterion for this to happen.

§2. Finite Fields

Let $\mathbb{F}_q$ denote a field that has a finite number q of elements in it. Clearly a finite field cannot have characteristic zero; so let p be the characteristic of $\mathbb{F}_q$. Then $\mathbb{F}_q$ contains the prime field $\mathbb{F}_p = \mathbb{Z}/p\mathbb{Z}$, and so is a vector space – necessarily finite dimensional – over $\mathbb{F}_p$. Let f denote its dimension as an $\mathbb{F}_p$-vector space. By choosing a basis, we can set up a 1-to-1 correspondence between the elements of this f-dimensional vector space and the set of all f-tuples of elements in $\mathbb{F}_p$. It follows that there must be p^f elements in $\mathbb{F}_q$. That is, q is a power of the characteristic p.

We shall soon see that for every prime power $q = p^f$ there is a field of q elements, and it is unique (up to isomorphism).

But first we investigate the multiplicative *order* of nonzero elements of $\mathbb{F}_q$. By the "order" of a nonzero element we mean the least positive power which is 1.

2.1 Existence of Multiplicative Generators of Finite Fields

There are $q - 1$ nonzero elements, and, by the definition of a field, they form an *abelian group* with respect to multiplication. This means that the product of two nonzero elements is nonzero, the associative law and commutative law hold, there is an identity element 1, and any nonzero element has an inverse. The group of nonzero elements of $\mathbb{F}_q$ is denoted $\mathbb{F}_q^*$.

It is an easily proved fact about finite groups that the order of any element must divide the number of elements in the group. Thus, the order of any $a \in \mathbb{F}_q^*$ divides $q - 1$.

Definition 2.1. A *generator* g of a finite field $\mathbb{F}_q$ is an element of order $q - 1$; equivalently, g is a generator if the powers of g run through all nonzero elements of $\mathbb{F}_q$.

The next theorem gives a basic fact about finite fields. It says that the nonzero elements of any finite field form a *cyclic group*; in other words, they are all powers of a single element.

Theorem 2.1. *Every finite field has a generator. If g is a generator of $\mathbb{F}_q^*$, then g^j is also a generator if and only if* g.c.d.$(j, q - 1) = 1$. *Thus, there are a total of $\varphi(q - 1)$ different generators of $\mathbb{F}_q^*$, where φ denotes the Euler φ-function.*

Proof. Suppose that $a \in \mathbb{F}_q^*$ has order d; that is, $a^d = 1$ and no lower power of a gives 1. As mentioned above, d divides $q - 1$. Since a^d is the smallest power that equals 1, it follows that the elements $a, a^2, \ldots, a^d = 1$ are distinct. We claim that the elements of order d are precisely the $\varphi(d)$ values a^j for which g.c.d.$(j, d) = 1$. First, since the d distinct powers of a all satisfy the equation $X^d = 1$ and since the polynomial $X^d - 1$ has at most d roots in a field, it follows that the powers of a exhaust all of the roots of this equation. Any element of order d must therefore be among the powers of a. However, not all powers of a have order d. Namely, if g.c.d.$(j, d) = d' > 1$, then a^j has lower order – in fact, $(a^j)^{d/d'} = 1$. Conversely, if g.c.d.$(j, d) = 1$, then we can write $ju - dv = 1$ for some integers u and v, and from this it follows that $a = a^{1+dv} = (a^j)^u$ is a power of a^j, and so a and a^j have the same order (because each is a power of the other). Thus, a^j has order d if and only if g.c.d.$(j, d) = 1$.

This means that, if there is any element a of order d, then there are exactly $\varphi(d)$ elements of order d. So for every d dividing $q-1$ there are only two possibilities: *no* element has order d, or exactly $\varphi(d)$ elements have order d. The rest of the argument depends on the following lemma.

Lemma 2.1. *For any integer $N > 1$ one has*

$$\sum_{d|N} \varphi(d) = N \ .$$

To prove the lemma, we partition the set $\{0, 1, \ldots, N-1\}$ according to g.c.d. with N. That is, we put j in the set S_d if g.c.d.$(j, N) = d$. As d ranges over all divisors of N, so does the integer d' determined by setting $N = d \cdot d'$. Clearly, the set S_d consists of the $\varphi(d')$ different values of $j = j'd$ for which g.c.d.$(j', d') = 1$. Since the set $\{0, 1, \ldots, N-1\}$ is the disjoint union of the S_d as d (and hence d') ranges over the divisors of N, it follows that $N = \sum_{d'|N} \varphi(d')$, as claimed.

We now return to the proof of the theorem. Every element of $\mathbb{F}_q^*$ has some order d dividing $q-1$. And there are either 0 or $\varphi(d)$ elements of each possible order d. Letting $N = q - 1$ in Lemma 2.1, we have $\sum_{d|q-1} \varphi(d) = q - 1$, which is the number of elements in $\mathbb{F}_q^*$. Thus, if we partition the elements of $\mathbb{F}_q^*$ according to their orders, we see that the only way that every element can have some order d dividing $q - 1$ is if there are always $\varphi(d)$ (and never 0) elements of order d. In particular, there are $\varphi(q - 1)$ elements of order $q - 1$; and, as we saw in the first paragraph of the proof, if g is any element of order $q - 1$, then the other elements of order $q - 1$ are precisely the powers g^j for which g.c.d.$(j, q - 1) = 1$. This completes the proof. $\square$

Corollary 2.1. *For every prime p, there exists an integer g such that the powers of g exhaust all nonzero residue classes modulo p.*

Example 2.1. We can get all residues mod 19 from 1 to 18 by taking powers of 2. Namely, the successive powers of 2 reduced mod 19 are: 2, 4, 8, 16, 13, 7, 14, 9, 18, 17, 15, 11, 3, 6, 12, 5, 10, 1.

2.2 Existence and Uniqueness of Finite Fields with Prime Power Number of Elements

We prove both existence and uniqueness by showing that a finite field of $q = p^f$ elements is the splitting field (see Definition 1.7) of the polynomial $X^q - X$. The following theorem shows that for every prime power q there is one and (up to isomorphism) only one finite field with q elements.

Theorem 2.2. *If $\mathbb{F}_q$ is a field of $q = p^f$ elements, then every element satisfies the equation $X^q - X = 0$, and $\mathbb{F}_q$ is precisely the set of roots of that equation. Conversely, for every prime power $q = p^f$ the splitting field over $\mathbb{F}_p$ of the polynomial $X^q - X$ is a field of q elements.*

Proof. First suppose that $\mathbb{F}_q$ is a finite field. Since the order of any nonzero element divides $q - 1$, it follows that any nonzero element satisfies the equation $X^{q-1} = 1$, and hence, if we multiply both sides by X, the equation $X^q = X$. Of course, the element 0 also satisfies the latter equation. Thus, all q elements of $\mathbb{F}_q$ are roots of the degree-q polynomial $X^q - X$. Since this polynomial cannot have more than q roots, its roots are precisely the elements of $\mathbb{F}_q$. Notice that this means that $\mathbb{F}_q$ is the splitting field of the polynomial $X^q - X$, that is, the smallest field extension of $\mathbb{F}_p$ that contains all of the roots of this polynomial.

Conversely, let $q = p^f$ be a prime power, and let $\mathbb{F}$ be the splitting field over $\mathbb{F}_p$ of the polynomial $X^q - X$. Note that $X^q - X$ has derivative $qX^{q-1} - 1 = -1$ (because the integer q is a multiple of p and so is zero in the field $\mathbb{F}_p$); hence, the polynomial $X^q - X$ has no common roots with its derivative (which has no roots at all), and therefore has no multiple roots. Thus, $\mathbb{F}$ must contain at least the q distinct roots of $X^q - X$. But we claim that the set of q roots is already a field. The key point is that a sum or product of two roots is again a root. Namely, if a and b satisfy the polynomial, we have $a^q = a$, $b^q = b$, and hence $(ab)^q = ab$, and so the product is also a root. To see that the sum $a + b$ also satisfies the polynomial $X^q - X = 0$, we note a fundamental fact about any field of characteristic p:

Lemma 2.2. $(a + b)^p = a^p + b^p$ in any field of characteristic p.

The lemma is proved by observing that all of the intermediate terms vanish in the binomial expansion $\sum_{j=0}^{p} \binom{p}{j} a^{p-j} b^j$, because $p!/(p - j)!j!$ is divisible by p for $0 < j < p$.

Repeated application of the lemma gives us: $a^p + b^p = (a + b)^p$, $a^{p^2} + b^{p^2} = (a^p + b^p)^p = (a + b)^{p^2}, \ldots, a^q + b^q = (a + b)^q$. Thus, if $a^q = a$ and $b^q = b$ it follows that $(a + b)^q = a + b$, and so $a + b$ is also a root of $X^q - X$. We conclude that the set of q roots is the smallest field containing the roots of $X^q - X$; in other words, the splitting field of this polynomial is a field of q elements. This completes the proof. $\square$

In the proof we showed that raising to the p-th power preserves addition and multiplication. We derive another important consequence of this in the next theorem.

Theorem 2.3. *Let $\mathbb{F}_q$ be the finite field of $q = p^f$ elements, and let σ be the map that sends every element to its p-th power: $\sigma(a) = a^p$. Then σ is an automorphism of the field $\mathbb{F}_q$ (a 1-to-1 map of the field to itself which preserves addition and multiplication – see Definition 1.5). The elements of $\mathbb{F}_q$ which are kept fixed by σ are precisely the elements of the prime field $\mathbb{F}_p$. The f-th power (and no lower power) of the map σ is the identity map.*

Proof. A map that raises to a power always preserves multiplication. The fact that σ preserves addition comes from Lemma 2.2. Notice that for any j the j-th power of σ (the result of applying σ repeatedly j times) is the map $a \mapsto a^{p^j}$. Thus, the elements left fixed by σ^j are the roots of $X^{p^j} - X$. If $j = 1$, these are precisely the p elements of the prime field (this is Fermat's Little Theorem). The elements left fixed by σ^f are the roots of $X^q - X$, i.e., all of $\mathbb{F}_q$. Since the f-th power of σ is the identity map, σ must be 1-to-1 (its inverse map is $\sigma^{f-1} : a \mapsto a^{p^{f-1}}$). No lower power of σ gives the identity map, since for $j < f$ not all of the elements of $\mathbb{F}_q$ could be roots of the polynomial $X^{p^j} - X$. This completes the proof. $\square$

Theorem 2.4. *In the notation of Theorem 2.3, if α is any element of $\mathbb{F}_q$, then the conjugates of α over $\mathbb{F}_p$ (the elements of $\mathbb{F}_q$ which satisfy the same monic irreducible polynomial with coefficients in $\mathbb{F}_p$) are the elements $\sigma^j(\alpha) = \alpha^{p^j}$.*

Proof. Let d be the degree of $\mathbb{F}_p(\alpha)$ as an extension of $\mathbb{F}_p$. That is, $\mathbb{F}_p(\alpha)$ is a copy of $\mathbb{F}_{p^d}$. Then α satifies $X^{p^d} - X$ but does not satisfy $X^{p^j} - X$ for any $j < d$. Thus, one obtains d distinct elements by repeatedly applying σ to α. It now suffices to show that each of these elements satisfies the same monic irreducible polynomial $f(X)$ that α does, in which case they must be the d roots. To do this, it is enough to prove that, if α satisfies a polynomial $f(X) \in \mathbb{F}_p[X]$, then so does α^p. Let $f(X) = \sum a_j X^j$, where $a_j \in \mathbb{F}_p$. Then $0 = f(\alpha) = \sum a_j \alpha^j$. Raising both sides to the p-th power gives $0 = \sum (a_j \alpha^j)^p$ (where we use Lemma 2.2). But $a_j^p = a_j$, by Fermat's Little Theorem, and so we have: $0 = \sum a_j (\alpha^p)^j = f(\alpha^p)$, as desired. This completes the proof. $\square$

2.3 Explicit Construction

So far our discussion of finite fields has been rather theoretical. Our only practical experience has been with the finite fields of the form $\mathbb{F}_p = \mathbb{Z}/p\mathbb{Z}$. We now discuss how to work with finite extensions of $\mathbb{F}_p$. At this point we should recall how in the case of the rational numbers $\mathbb{Q}$ we work with an extension such as $\mathbb{Q}(\sqrt{2})$. Namely, we get this field by taking a root α of the equation $X^2 - 2$ and looking at expressions of the form $a + b\alpha$, which are added and multiplied in the usual way, except that α^2 should always be replaced by 2. (In the case of $\mathbb{Q}(\sqrt[3]{2})$ we work with expressions of the form $a + b\alpha + c\alpha^2$, and when we multiply we always replace α^3 by 2.) We can take the same general approach with finite fields.

Example 2.2. To construct $\mathbb{F}_9$ we take any monic quadratic polynomial in $\mathbb{F}_3[X]$ which has no roots in $\mathbb{F}_3$. By trying all possible choices of coefficients and testing whether the elements $0, \pm 1 \in \mathbb{F}_3$ are roots, we find that there are three monic irreducible quadratics: $X^2 + 1$, $X^2 \pm X - 1$. If, for example, we take α to be a root of $X^2 + 1$ (let's call it i rather than α – after all, we are simply adjoining a square root of -1), then the elements of $\mathbb{F}_9$ are all combinations $a + bi$, where a and b are 0, 1, or -1. Arithmetic in $\mathbb{F}_9$ is thus a lot like arithmetic in the Gaussian integers (the set of complex numbers $a + bi$ where a and b are integers), except that in $\mathbb{F}_9$ we work with coefficients a and b that are in the tiny field $\mathbb{F}_3$.

Notice that the element i that we adjoined is *not* a generator of $\mathbb{F}_9^*$, since it has order 4 rather than $q - 1 = 8$. If, however, we adjoin a root α of $X^2 - X - 1$, we can get all nonzero elements of $\mathbb{F}_9$ by taking the successive powers of α (remember that α^2 must always be replaced by $\alpha + 1$, since α satisfies $X^2 = X + 1$): $\alpha^1 = \alpha$,

$\alpha^2 = \alpha + 1$, $\alpha^3 = -\alpha + 1$, $\alpha^4 = -1$, $\alpha^5 = -\alpha$, $\alpha^6 = -\alpha - 1$, $\alpha^7 = \alpha - 1$, $\alpha^8 = 1$. We sometimes say that the polynomial $X^2 - X - 1$ is *primitive*, meaning that any root of the irreducible polynomial is a generator of the group of nonzero elements of the field. There are $4 = \varphi(8)$ generators of $\mathbb{F}_9^*$, by Theorem 2.1: two are the roots of $X^2 - X - 1$ and two are the roots of $X^2 + X - 1$. (The second root of $X^2 - X - 1$ is the conjugate of α, namely, $\sigma(\alpha) = \alpha^3 = -\alpha + 1$.) Of the remaining four nonzero elements, two are the roots of $X^2 + 1$ (namely $\pm i = \pm(\alpha + 1)$) and the other two are the two nonzero elements ± 1 of $\mathbb{F}_3$ (which are roots of the degree-1 monic irreducible polynomials $X - 1$ and $X + 1$).

Recall that in any finite field $\mathbb{F}_q$, $q = p^f$, each element α satisfies a unique monic polynomial over $\mathbb{F}_p$ of some degree d. Then the field $\mathbb{F}_p(\alpha)$ obtained by adjoining this element to the prime field is an extension of degree d that is contained in $\mathbb{F}_q$. That is, it is a copy of the field $\mathbb{F}_{p^d}$. Since the big field $\mathbb{F}_{p^f}$ contains $\mathbb{F}_{p^d}$, and so is an $\mathbb{F}_{p^d}$–vector space of some dimension f', it follows that the number of elements in $\mathbb{F}_{p^f}$ must be $(p^d)^{f'}$; in other words, $f = df'$. Thus, $d | f$. Conversely, for any $d | f$ the finite field $\mathbb{F}_{p^d}$ is contained in $\mathbb{F}_q$, because any solution of $X^{p^d} = X$ is also a solution of $X^{p^f} = X$. (To see this, note that for any d', if you repeatedly replace X by X^{p^d} on the left in the equation $X^{p^d} = X$, you can obtain $X^{p^{dd'}} = 1$.) Thus, we have proved:

Theorem 2.5. *The subfields of $\mathbb{F}_{p^f}$ are the $\mathbb{F}_{p^d}$ for d dividing f. If an element of $\mathbb{F}_{p^f}$ is adjoined to $\mathbb{F}_p$, one obtains one of these fields.*

It is now easy to prove a formula that is useful in determining the number of irreducible polynomials of a given degree.

Theorem 2.6. *For any $q = p^f$ the polynomial $X^q - X$ factors in $\mathbb{F}_p[X]$ into the product of all monic irreducible polynomials of degrees d dividing f.*

Proof. If we adjoin to $\mathbb{F}_p$ a root α of any monic irreducible polynomial of degree $d | f$, we obtain a copy of $\mathbb{F}_{p^d}$, which is contained in $\mathbb{F}_{p^f}$. Since α then satisfies $X^q - X = 0$, the monic irreducible must divide that polynomial. Conversely, let $f(X)$ be a monic irreducible polynomial that divides $X^q - X$. Then $f(X)$ must have its roots in $\mathbb{F}_q$ (since that's where all of the roots of $X^q - X$ are). Thus $f(X)$ must have degree dividing f, by Theorem 2.5, since adjoining a root gives a subfield of $\mathbb{F}_q$. Thus, the monic irreducible polynomials that divide $X^q - X$ are precisely all of the ones of degree dividing f. Since we saw that $X^q - X$ has no multiple factors, this means that $X^q - X$ is equal to the product of all such irreducible polynomials, as was to be proved. $\square$

Corollary 2.2. *If f is a prime number, then there are $(p^f - p)/f$ distinct monic irreducible polynomials of degree f in $\mathbb{F}_p[X]$.*

Notice that $(p^f - p)/f$ is an integer because of Fermat's Little Theorem for the prime f, which states that $p^f \equiv p \bmod f$. To prove the corollary, let n be the number of monic irreducible polynomials of degree f. According to the proposition, the degree-p^f polynomial $X^{p^f} - X$ is the product of n polynomials of degree f and the p degree-1 irreducible polynomials $X - a$ for $a \in \mathbb{F}_p$. Thus, equating degrees gives: $p^f = nf + p$, from which the desired equality follows.

More generally, suppose that f is not necessarily prime. Let n_d denote the number of monic irreducible polynomials of degree d over $\mathbb{F}_p$. Using an argument similar to the proof of Corollary 2.2, we find that $n_f = (p^f - \sum d\, n_d)/f$, where the summation is over all $d < f$ that divide f.

We now extend the time estimates in Chapter 2 for arithmetic modulo p to general finite fields.

Theorem 2.7. *Let $\mathbb{F}_q$, where $q = p^f$, be a finite field, and let $F(X)$ be an irreducible polynomial of degree f over $\mathbb{F}_p$. Then two elements of $\mathbb{F}_q$ can be multiplied or divided in $O(\ln^2 q)$ bit operations. If N is a positive integer, then an element of $\mathbb{F}_q$ can be raised to the N-th power in $O(\ln N \ln^2 q)$ bit operations.*

Proof. An element of $\mathbb{F}_q$ is a polynomial with coefficients in $\mathbb{F}_p = \mathbb{Z}/p\mathbb{Z}$ regarded modulo $F(X)$. To multiply two such elements, we multiply the polynomials – this requires $O(f^2)$ multiplications of integers modulo p (and some additions of integers modulo p, which take much less time) – and then divide the polynomial $F(X)$ into the product, taking the remainder polynomial as our answer. The polynomial division involves $O(f)$ divisions of integers modulo p and $O(f^2)$ multiplications of integers modulo p. Since a multiplication modulo p takes $O(\ln^2 p)$ bit operations, and a division (using the Euclidean algorithm, for example) takes $O(\ln^2 p)$ bit operations (see Example 3.4 of Chapter 2), the total number of bit operations is: $O(f^2 \ln^2 p) = O(\ln^2 q)$. To prove the same result for division, it suffices to show that the reciprocal of an element can be found in time $O(\ln^2 q)$. Using the Euclidean algorithm for polynomials over the field $\mathbb{F}_p$ (see §3 below), we must write 1 as a linear combination of our given element in $\mathbb{F}_q$ (that is, a given polynomial of degree $< f$) and the fixed degree-f polynomial $F(X)$. Using the same argument as in Example 3.4 of Chapter 2, we see that this involves $O(f^2)$ operations in $\mathbb{F}_p$, and hence $O(f^2 \ln^2 p) = O(\ln^2 q)$ bit operations. Finally, an N-th power can be computed by the repeated squaring method in the same way as modular exponentiation (see Example 3.5 of Chapter 2). This takes $O(\ln N)$ multiplications (or squarings) of elements of $\mathbb{F}_q$, and hence $O(\ln N \ln^2 q)$ bit operations. This completes the proof. $\square$

Exercises for § 2

1. For $p = 2, 3, 5, 7, 11, 13$ and 17, find the smallest positive integer which generates $\mathbb{F}_p^*$, and determine how many of the integers $1, 2, 3, \ldots, p-1$ are generators.

2. Let $(\mathbb{Z}/p^\alpha\mathbb{Z})^*$ denote all residues modulo p^α that are *invertible*, i.e., are not divisible by p. **Warning:** Be sure not to confuse $\mathbb{Z}/p^\alpha\mathbb{Z}$ (which has $p^\alpha - p^{\alpha-1}$ invertible elements) with $\mathbb{F}_{p^\alpha}$ (in which all elements except 0 are invertible). The two are the same only when $\alpha = 1$.

(a) Let $p > 2$, and let g be an integer that generates $\mathbb{F}_p^*$. Let α be any integer greater than 1. Prove that either g or $(p+1)g$ generates $(\mathbb{Z}/p^\alpha\mathbb{Z})^*$. Thus, the latter is also a *cyclic group*.

(b) Prove that if $\alpha > 2$, then $(\mathbb{Z}/2^\alpha\mathbb{Z})^*$ is *not* cyclic, but that the number 5 generates a *subgroup* consisting of half of its elements, namely those which are $\equiv 1 \bmod 4$.

3. If p is a prime not equal to 7, find a simple way to find the degree over $\mathbb{F}_p$ of the splitting field of the polynomial $X^6 + X^5 + X^4 + X^3 + X^2 + X + 1$.

4. For each degree $d \leq 6$, find the number of irreducible polynomials over $\mathbb{F}_2$ of degree d, and make a list of them.

5. For each degree $d \leq 6$, find the number of monic irreducible polynomials over $\mathbb{F}_3$ of degree d, and for $d \leq 3$ make a list of them.

6. Suppose that f is a power of a prime ℓ. Find a simple formula for the number of monic irreducible polynomials of degree f over $\mathbb{F}_p$.

7. Suppose that $\alpha \in \mathbb{F}_{p^2}$ satisfies the polynomial $X^2 + aX + b$, where $a, b \in \mathbb{F}_p$.
(a) Prove that α^p also satisfies this polynomial.
(b) Prove that if $\alpha \notin \mathbb{F}_p$, then $a = -\alpha - \alpha^p$ and $b = \alpha^{p+1}$.
(c) Prove that if $\alpha \notin \mathbb{F}_p$ and $c, d \in \mathbb{F}_p$, then $(c\alpha + d)^{p+1} = d^2 - acd + bc^2$ (which is an element of $\mathbb{F}_p$).
(d) Let i be a square root of -1 in $\mathbb{F}_{19^2}$. Use part (c) to find $(2 + 3i)^{101}$ (that is, write it in the form $a + bi$, $a, b \in \mathbb{F}_{19}$).

8. For each of the following fields $\mathbb{F}_q$, where $q = p^f$, find an irreducible polynomial with coefficients in the prime field whose root α is primitive (i.e., generates $\mathbb{F}_q^*$), and write all of the powers of α as polynomials in α of degree less than f: (a) $\mathbb{F}_4$; (b) $\mathbb{F}_8$; (c) $\mathbb{F}_{27}$; (d) $\mathbb{F}_{25}$.

9. (a) Under what conditions on p and f is *every* element of $\mathbb{F}_{p^f}$ besides 0 and 1 a generator of $\mathbb{F}_{p^f}^*$?
(b) Under what conditions is every element besides 0 and 1 either a generator or the square of a generator?

10. Let σ be the automorphism of $\mathbb{F}_q$ in Theorem 4.2. Prove that the set of elements left fixed by σ^j is the field $\mathbb{F}_{p^d}$, where $d = $ g.c.d.(j, f).

11. Prove that if b is a generator of $\mathbb{F}_{p^f}^*$ and if $d|f$, then $b^{(p^f-1)/(p^d-1)}$ is a generator of $\mathbb{F}_{p^d}^*$.

12. Prove that the number of k-th roots of unity in $\mathbb{F}_{p^f}$ is equal to g.c.d.$(k, p^f - 1)$.

13. Let $q = p^f$, and consider the field $\mathbb{K} = \mathbb{F}_{p^{fn}} = \mathbb{F}_{q^n}$. Prove that the polynomial $g(X) = \sum_{i=0}^{n-1} X^{q^i}$ gives an $\mathbb{F}_q$-linear map from $\mathbb{K}$ to $\mathbb{F}_q$ that is surjective, i.e., takes all possible values in $\mathbb{F}_q$. This polynomial is called the "trace". Also show that there are exactly q^{n-1} elements of $\mathbb{K}$ with each possible trace. In other words, the equation $g(X) = y$ has q^{n-1} solutions in $\mathbb{K}$ for each $y \in \mathbb{F}_q$.

14. Let q be a power of an odd prime.

(a) Prove that an element $z \in \mathbb{F}_q^*$ has a square root in $\mathbb{F}_q^*$ if and only if $z^{(q-1)/2} = 1$. If $q \equiv 1 \pmod 4$, and if $z \in \mathbb{F}_q^*$ is chosen at random, show that there is a 50% chance that $y = z^{(q-1)/4}$ is a square root of -1.

(b) Describe how the Euclidean algorithm (see §3.3 of Chapter 2) works in the Gaussian integers.

(c) Suppose that $p \equiv 1 \pmod 4$ and y is an integer between 1 and p such that $y^2 \equiv -1 \pmod p$. Show that by applying the Euclidean algorithm in part (b) to the Gaussian integers $y + i$ and p, one can write p as a sum of two squares: $p = c^2 + d^2$. For example, carry this out when $p = 29$ and $y = 12$. If p is a very large prime, then the above procedure is an efficient way to write p as a sum of two squares.

§ 3. The Euclidean Algorithm for Polynomials

In this section we are still working with polynomials in a single variable. Multivariable polynomials will be the subject of §§4–5.

Definition 3.1. The *greatest common divisor* of two polynomials $f, g \in \mathbb{F}[X]$ is the monic polynomial of largest degree that divides them both. Equivalently, it is the unique monic polynomial that divides f and g and is divisible by any other polynomial dividing f and g.

As in the case of integers, we find the g.c.d. of two polynomials by means of the Euclidean algorithm. The Euclidean algorithm for polynomials over a field $\mathbb{F}$ is very similar to the Euclidean algorithm for integers (see Example 3.4 of Chapter 2). Here is an example over the field $\mathbb{F}_2$, where the calculations are particularly efficient because the field operations are trivial.

Example 3.1. Let $f(X) = X^4 + X^3 + X^2 + 1$, $g = X^3 + 1 \in \mathbb{F}_2[X]$. Find g.c.d.$(f, g)$ using the Euclidean algorithm for polynomials, and express the g.c.d. in the form $u(X)f(X) + v(X)g(X)$.

Solution. Polynomial division gives us the sequence of equalities below, which lead to the conclusion that g.c.d.$(f, g) = X + 1$. (Of course, in a field of characteristic 2 adding is the same as subtracting; that is, $a - b = a + b - 2b = a + b$.) We have:

$$f = (X + 1)g + (X^2 + X)$$
$$g = (X + 1)(X^2 + X) + (X + 1)$$
$$X^2 + X = X(X + 1) .$$

If we now work backwards in the above column of equalities, we can express $X + 1$ as a linear combination of f and g:

$$X + 1 = g + (X + 1)(X^2 + X)$$
$$= g + (X + 1)(f + (X + 1)g)$$
$$= (X + 1)f + (X^2)g .$$

We now break up the Euclidean algorithm into smaller steps, so as to get a closer look at the procedure with an eye toward generalizing it to polynomials in several variables (see §5). At each stage we determine just one term of a polynomial division. We shall give an example over a bigger field.

Example 3.2. Let $f(X) = X^3 - 2X^2 + 5$, $g(X) = 2X^2 + 3X - 4 \in \mathbb{F}_{11}[X]$. Show that $f(X)$ and $g(X)$ are relatively prime, and find $u(X)$ and $v(X)$ such that $uf + vg = 1$.*

Solution.

$$f(X) = X^3 - 2X^2 + 5 = (-5X)g(X) + (2X^2 + 2X + 5)$$
$$= (-5X + 1)g(X) + (-X - 2)$$
$$g(X) = 2X^2 + 3X - 4 = (-2X)(-X - 2) + (-X - 4)$$
$$= (-2X + 1)(-X - 2) + (-2) .$$

The last step in a g.c.d. computation is to divide by a suitable constant (here it is -2) to get a monic polynomial (in this case the constant 1).

We omit the details of the computation of $u(X)$ and $v(X)$ such that $uf + vg = 1$. One obtains $u(X) = -X - 5$, $v(X) = -5X^2 - 2X - 1$.

Exercises for §3

1. Use the polynomial version of the Euclidean algorithm to find $d =$g.c.d.(f, g) for $f, g \in \mathbb{F}_p[X]$ in each of the following examples. In each case express $d(X)$ in the form $d(X) = u(X)f(X) + v(X)g(X)$.
(a) $f = X^3 + X + 1$, $g = X^2 + X + 1$, $p = 2$;
(b) $f = X^6 + X^5 + X^4 + X^3 + X^2 + X + 1$, $g = X^4 + X^2 + X + 1$, $p = 2$;
(c) $f = X^3 - X + 1$, $g = X^2 + 1$, $p = 3$;
(d) $f = X^5 + X^4 + X^3 - X^2 - X + 1$, $g = X^3 + X^2 + X + 1$, $p = 3$;
(e) $f = X^5 + 88x^4 + 73X^3 + 83X^2 + 51X + 67$, $g = X^3 + 97X^2 + 40X + 38$, $p = 101$.

2. By computing g.c.d.(f, f'), find all multiple roots of $f(X) = X^7 + X^5 + X^4 - X^3 - X^2 - X + 1 \in \mathbb{F}_3[X]$ in its splitting field.

3. State and prove a polynomial analogue of the Chinese Remainder Theorem (see Exercise 9 in §3 of Chapter 2).

---

* For no special reason we are using the least absolute representatives modulo 11 – the numbers $\{0, \pm 1, \pm 2, \pm 3, \pm 4, \pm 5\}$ – rather than the least nonnegative residues.

§ 4. Polynomial Rings

4.1 Basic Definitions

Definition 4.1. A *ring** is a set **R** with a *multiplication* operation and an *addition* operation that satisfy all of the rules of a field (see Definition 1.1), except that nonzero elements need not have multiplicative inverses.

The most familiar example of Definition 4.1 is the ring of integers $\mathbb{Z}$. Another example is the ring of Gaussian integers $\mathbb{Z}[i]$; it consists of all complex numbers of the form $a + bi$, where a and b are integers. A third example of a ring is the set of all expressions of the form $a + bX$, where a and b are in a field $\mathbb{F}$ and where multiplication is defined by the rule $(a + bX)(a' + b'X) = aa' + (ab' + a'b)X$. Also note that any field $\mathbb{F}$ is automatically a ring.

Definition 4.2. An *integral domain* is a ring **R** with no nontrivial zero divisors. This means that if $xy = 0$ for $x, y \in \mathbf{R}$, then either x or y is zero.

The first two examples above – $\mathbb{Z}$ and $\mathbb{Z}[i]$ – are integral domains, but the third one is not. Namely, in the third example $(bX)(b'X) = 0$ for $b, b' \in \mathbb{F}$.

Definition 4.3. If **R** is a ring, then the *polynomial ring* in m variables $X = \{X_1, \ldots, X_m\}$ over **R** is the set of all finite expressions of the form $\sum a_{i_1, \ldots, i_m} X_1^{i_1} \cdots X_m^{i_m}$, where $a_{i_1, \ldots, i_m} \in \mathbf{R}$ and the i_j are nonnegative integers. Such polynomials are added and multiplied in the usual way. The polynomial ring is denoted $\mathbf{R}[X]$ or else $\mathbf{R}[X_1, \ldots, X_m]$. We sometimes use the vector notation $\mathbf{i} = (i_1, \ldots, i_m)$ and write $a_\mathbf{i} X^\mathbf{i}$ to denote $a_{i_1, \ldots, i_m} X_1^{i_1} \cdots X_m^{i_m}$. The *total degree* of a monomial term $a_\mathbf{i} X^\mathbf{i}$ is defined to be $i_1 + \cdots + i_m$, and the *total degree* of a polynomial is the maximum of the total degrees of its nonzero monomial terms.

Definition 4.4. If **R** is a ring, then an *ideal* of **R** is a subset that is closed under addition and subtraction and under multiplication by any element of **R**. That is, an additive subgroup $I \subset \mathbf{R}$ is an ideal if $ra \in I$ for every $a \in I$ and every $r \in \mathbf{R}$. The *unit ideal* is the ideal consisting of all elements of **R**; a *proper* ideal is an ideal that is not the unit ideal, that is, an ideal that is properly contained in **R**. Obviously, an ideal is the unit ideal if and only if it contains 1. By a *nontrivial* ideal we mean any ideal other than the zero ideal or the unit ideal.

Notice that if **R** is a field, then it has no nontrivial ideals. This is because, if $a \in I$ is a nonzero element, then a has an inverse a^{-1} in **R**, and so $1 = a^{-1}a$ is in I. Conversely, if **R** is not a field, then there is some nonzero element $r \in \mathbf{R}$ that does not have a multiplicative inverse. Then the ideal of all multiplies of r – this is denoted $\mathbf{R}r$ – is a nonzero proper ideal.

---

* More precisely, a commutative ring with identity. However, all rings in this book are commutative and contain 1; so we shall simply use the term "ring" for a commutative ring with identity.

Definition 4.5. A proper ideal I of **R** is said to be a *maximal* ideal if there is no ideal of **R** strictly contained between I and **R**. A proper ideal I of **R** is said to be a *prime* ideal if, whenever a product $r_1 r_2$ belongs to I (where $r_1, r_2 \in$ **R**) either r_1 or r_2 must belong to I.

In the case of the familiar ring $\mathbb{Z}$ it is not hard to show that any maximal ideal is the set of multiples of some fixed prime number. The prime ideals are the same, except that the zero ideal is a prime ideal as well (but it is not a maximal ideal).

In the case of the polynomial ring in two variables $\mathbb{C}[X, Y]$ it is not hard to show that any maximal ideal is of the following form: it is the set of all polynomials that vanish at a fixed point $(x_0, y_0) \in \mathbb{C}^2$. The prime ideals consist of (1) all of the maximal ideals, (2) the zero ideal, and (3) all ideals of the form $\mathbb{C}[X, Y] f(X, Y)$, where $f(X, Y)$ is a fixed irreducible polynomial.

Definition 4.6. A *set of generators* of an ideal I in a ring **R** is a set of elements of I such that any element of I is a finite linear combination of elements in the set (with coefficients in **R**). An ideal is said to be *finitely generated* if it has a finite set of generators. If I is generated by the set of elements $\{f_1, \ldots, f_l\} \subset I$, then we write either $I = \sum_{i=1}^{l} \mathbf{R} f_i$ or else simply $I = (f_1, \ldots, f_l)$.

Definition 4.7. A *principal* ideal I in a ring **R** is an ideal generated by a single element. That is, for some fixed $f \in I$ the ideal I consists of all elements of **R** of the form af, where $a \in$ **R**. An integral domain is said to be a *principal ideal domain* (or PID) if all of its ideals are principal ideals.

For example, $\mathbb{Z}$ is a PID. That is, any ideal $I \subset \mathbb{Z}$ is of the form $\mathbb{Z}a$ for some integer a. If we are given a set of generators of I, this number a is the g.c.d. of these generators; we find a using the Euclidean algorithm (see Example 3.4 of Chapter 2).

It is also not hard to show that $\mathbb{Z}[i]$ is a PID.

4.2 The Hilbert Basis Theorem

Definition 4.8. A ring **R** is said to be *Noetherian* if any ideal of **R** is finitely generated.

Any field (or any PID) is trivially a Noetherian ring, since every ideal has a single generator. An important and less trivial class of examples is the polynomial rings. The Hilbert Basis Theorem essentially says that all such rings are Noetherian.

Theorem 4.1. *If* **R** *is a Noetherian ring, then so is the polynomial ring in one variable* **R**$[X]$.

Proof. Let I be an ideal of **R**$[X]$. We must show that I is finitely generated. For each $n = 0, 1, 2, \ldots$ let $J_n \subset$ **R** denote the set consisting of 0 and all leading coefficients of polynomials in I of degree n. It is easy to check that J_n is an ideal of **R**. It is also clear that $J_n \subset J_{n+1}$ for $n = 0, 1, 2, \ldots$; this is because, if the degree-n polynomial f is in I, then so is the degree-$(n+1)$ polynomial Xf, which has the same leading term. We set $J = \bigcup_{n=0}^{\infty} J_n$. Clearly J is an ideal of $\mathbf{R}$. By assumption, any ideal of $\mathbf{R}$ is finitely generated. That means that J is generated by a finite set of elements r_i, each of which is in some J_{n_i}. If we take N to be the maximum n_i, we conclude that the entire generating set – and hence all of J – is contained in J_N. That is, $J = J_N$. (In other words, $J_{n+1} = J_n$ for $n \geq N$.)

Since $\mathbf{R}$ is assumed to be Noetherian, each of the ideals J_n has a finite set of generators $\{r_{n,1}, \ldots, r_{n,l_n}\}$. The union of these sets as $n = 0, 1, \ldots, N$ generates all of J. For each $r_{n,i}$ let $f_{n,i}$ denote a degree-n polynomial in I whose leading coefficient is $r_{n,i}$. We claim that the union of the sets $\{f_{n,1}, \ldots, f_{n,l_n}\}$ as $n = 0, 1, \ldots, N$ generates all of I.

To see this, let us suppose that $f \in I$ has degree n. Its leading term, which belongs to J_n, can be written in the form $\sum_i a_i r_{n,i}$ with $a_i \in \mathbf{R}$, provided that $n \leq N$. If $n > N$, then $J_n = J_N$, and we can write the leading term as a linear combination of the $r_{N,i}$. This means that the polynomial $f - \sum_i a_i f_{n,i}$ (or $f - \sum_i a_i f_{N,i}$ in the case $n > N$) is an element of I of degree less than n. In other words, f can be expressed as a linear combination of the $f_{n,i}$ plus a polynomial $\tilde{f} \in I$ of lower degree. If we then apply the same argument to $\tilde{f}$, that is, if we express it as a linear combination of polynomials in our set plus a polynomial of still lower degree, and if we continue in this way, we eventually arrive at an expression for f as a linear combination of $f_{n,i}$, $n = 0, 1, \ldots, N$, $i = 1, 2, \ldots, l_n$. This completes the proof of the theorem. $\square$

Corollary 4.1. *If $\mathbb{F}$ is a field and $X = \{X_1, \ldots, X_m\}$ is a finite set of variables, then $\mathbb{F}[X]$ is a Noetherian ring.*

Proof. The corollary follows immediately from the theorem if we use induction on m. $\square$

Corollary 4.2. *There is no infinite sequence of strictly increasing ideals $I_1 \subset I_2 \subset \cdots$ in $\mathbb{F}[X_1, \ldots, X_m]$.*

Proof. If there were, let I denote the union of all of the ideals, which itself is clearly an ideal. By Corollary 4.1, I has a finite set of generating elements. Each of them is an element of some I_i. Let n be the maximum of these i. Then $I = I_n$, and so we could not have a strictly larger ideal $I_{n+1} \subset I$. $\square$

4.3 Homomorphisms and Transcendental Elements

Definition 4.9. If $\mathbf{R}$ and $\mathbf{R}'$ are two rings, then a *homomorphism* φ from $\mathbf{R}$ to $\mathbf{R}'$ is a map that preserves addition and multiplication and takes the multiplicative identity $1 \in \mathbf{R}$ to the multiplicative identity in $\mathbf{R}'$ (which is usually also denoted 1). That is,

$$\varphi(r_1 \pm r_2) = \varphi(r_1) \pm \varphi(r_2) \qquad \text{and} \qquad \varphi(r_1 r_2) = \varphi(r_1)\varphi(r_2)$$

for all $r_1, r_2 \in \mathbf{R}$, and $\varphi(1) = 1$.

Here are two basic examples of ring homomorphisms.

Example 4.1. Let I be an ideal of the ring $\mathbf{R}$, and define $\mathbf{R}'$ to be the set of equivalence classes of elements of $\mathbf{R}$, where $r_1 \sim r_2$ if and only if $r_2 - r_1 \in I$. We write the equivalence class of $r \in \mathbf{R}$ in the form $r + I \in \mathbf{R}'$. The multiplication and addition operations in $\mathbf{R}$ carry over to similar operations in $\mathbf{R}'$, so $\mathbf{R}'$ is a ring that is denoted $\mathbf{R}/I$ and is called the "quotient ring" of $\mathbf{R}$ by the ideal I. The map $r \mapsto r + I$ from $\mathbf{R}$ to $\mathbf{R}'$ is a ring homomorphism, called the "canonical surjection" from $\mathbf{R}$ to $\mathbf{R}/I$.

It is an easy consequence of Definition 4.5 that an ideal $I \subset \mathbf{R}$ is a maximal ideal if and only if R/I is a field, and it is a prime ideal if and only if R/I is an integral domain.

Example 4.2. Let $\mathbf{R}$ be a ring, and let $\mathbf{R}'$ be a ring containing $\mathbf{R}$. Let $X = \{X_1, \ldots, X_m\}$ be a finite set of variables. Given any m elements $t_1, \ldots, t_m \in \mathbf{R}'$ there is a unique ring homomorphism from the polynomial ring $\mathbf{R}[X]$ to $\mathbf{R}'$ such that X_i goes to t_i for $i = 1, \ldots, m$. The image of the homomorphism is a subring of $\mathbf{R}'$ containing $\mathbf{R}$ that is denoted $\mathbf{R}[t_1, \ldots, t_m]$.

In Example 4.2 suppose that $\mathbf{R} = \mathbb{F}$ and $\mathbf{R}' = \mathbb{F}'$ are fields. If all of the t_i are algebraic over $\mathbb{F}$ (that is, if each one satisfies a polynomial equation with coefficients in $\mathbb{F}$), then it is not hard to show that $\mathbb{F}[t_1, \ldots, t_m]$ is a finite dimensional vector space over $\mathbb{F}$, and so itself is a field. On the other hand, if any of the t_i is *transcendental* over $\mathbb{F}$ – that is, if t_i does not satisfy any polynomial equation over $\mathbb{F}$ – then one can show that $\mathbb{F}[t_1, \ldots, t_m]$ is *not* a field. We leave the proofs, which are not difficult and can be found in many textbooks, to the reader.

4.4 Hilbert Nullstellensatz*

Suppose that we have a set of polynomials $f_i \in \mathbb{F}[X] = \mathbb{F}[X_1, \ldots, X_m]$ with coefficients in a field $\mathbb{F}$. We might be interested in the set of points $(a_1, \ldots, a_m) \in \mathbb{F}^m$ where all of these polynomials vanish, that is, where $f_i(a_1, \ldots, a_m) = 0$ for all i. Let I be the ideal generated by all of the f_i: this is the set of all linear combinations of the f_i with coefficients in $\mathbb{F}[X]$. Clearly, if all of the f_i vanish at the point $(a_1, \ldots, a_m)$, then so do all $f \in I$.

Definition 4.10. Let $\mathbb{K}$ be an extension field of $\mathbb{F}$. The *zero set* of an ideal $I \subset \mathbb{F}[X_1, \ldots, X_m]$ in $\mathbb{K}^m$ is the set of all $(a_1, \ldots, a_m) \in \mathbb{K}^m$ such that $f(a_1, \ldots, a_m) = 0$ for all $f \in I$.

To what extent can we characterize an ideal of $\mathbb{F}[X]$ by giving its zero set in $\mathbb{F}^m$? In other words, if two ideals I and I' vanish at exactly the same subset of $\mathbb{F}^m$, are they the same ideal? The answer is no, as we see from the following examples.

Example 4.3. (a) Let $\mathbb{F}$ be the real numbers (or else the field of 3 elements, or any other field that does not contain a square root of -1), and let $m = 1$. Let I be

---

* The word "Nullstellensatz" is German for "zero point theorem".

the unit ideal and let I' be the ideal generated by the polynomial $X^2 + 1 \in \mathbb{F}[X]$. Both I and I' vanish at the empty set of points.

(b) Let $\mathbb{F} = \mathbb{F}_q$, and let $m = 1$. Let I be the zero ideal, and let I' be the ideal generated by the polynomial $X^q - X \in \mathbb{F}[X]$. Both I and I' vanish at all points of $\mathbb{F}$.

(c) Let $\mathbb{F}$ be any field, and let $m = 1$. Let I be the ideal generated by the polynomial $X \in \mathbb{F}[X]$, and let I' be the ideal generated by $X^2 \in \mathbb{F}[X]$. Both I and I' vanish on the set $\{0\}$.

Despite the possibilities illustrated in Example 4.3, the correspondence between ideals and their "zero sets" is crucial for the branch of mathematics known as *algebraic geometry*. A fundamental theorem of Hilbert clarifies what is going on in Example 4.3, and gives a more satisfactory answer to the question posed above.

The theorem that follows is known as "weak Hilbert Nullstellensatz".

Theorem 4.2. *Suppose that $\mathbb{F}$ is an algebraically closed field, and I is a proper ideal of the polynomial ring $\mathbb{F}[X] = \mathbb{F}[X_1, \ldots, X_m]$. Then there exists $a_1, \ldots, a_m \in \mathbb{F}$ such that all of the polynomials in I vanish at the point $(a_1, \ldots, a_m)$.*

Sketch of Proof. If I is not itself a maximal ideal, let M be a maximal ideal of $\mathbb{F}[X]$ that contains it. Let $\mathbb{F}'$ denote the quotient ring $\mathbb{F}[X]/M$, and let t_i denote the image of X_i under the canonical surjection from $\mathbb{F}[X]$ to $\mathbb{F}'$. By the remark following Example 4.1, $\mathbb{F}'$ is a field. By the remark following Example 4.2, all of the t_i are algebraic over $\mathbb{F}$. Since $\mathbb{F}$ is algebraically closed, this means that $t_i \in \mathbb{F}$, $i = 1, \ldots, m$. We take $(a_1, \ldots, a_m)$ to be the point $(t_1, \ldots, t_m)$. Then all of the polynomials $X_i - a_i$, $i = 1, \ldots, m$, are in M. Since the ideal generated by the $X_i - a_i$ is maximal, this ideal must be M. It follows that all polynomials in M – and hence all polynomials in I – vanish at $(a_1, \ldots, a_m)$. $\square$

The next theorem is known as "strong Hilbert Nullstellensatz".

Theorem 4.3. *Suppose that $\mathbb{F}$ is an algebraically closed field, and I is an ideal of the polynomial ring $\mathbb{F}[X] = \mathbb{F}[X_1, \ldots, X_m]$. Suppose that $f \in \mathbb{F}[X]$ vanishes at every point $(a_1, \ldots, a_m) \in \mathbb{F}^m$ at which all of the polynomials in I vanish. Then there exists an integer n such that $f^n \in I$.*[*]

Sketch of Proof. Let $f \in \mathbf{R} = \mathbb{F}[X]$ satisfy the condition in the theorem. We derive Theorem 4.3 from Theorem 4.2, which we apply to the polynomial ring $\mathbf{R}' = \mathbb{F}[X_1, \ldots, X_m, X_{m+1}]$ in one more variable. Let J be the ideal of $\mathbf{R}'$ that is generated by all polynomials in I and also the polynomial $1 - fX_{m+1}$. If $(a_1, \ldots, a_m, a_{m+1})$ were a point where all of the polynomials in J vanish, then all of the polynomials in I would vanish at $(a_1, \ldots, a_m)$, and so f would also vanish there. But then $1 - fX_{m+1}$ would take the value 1 at $(a_1, \ldots, a_m, a_{m+1})$. Hence, there is no point $(a_1, \ldots, a_m, a_{m+1})$ where all of the polynomials in J vanish. By

---

[*] The set of all $f \in \mathbb{F}[X]$ such that $f^n \in I$ for some integer n is called the *radical* of I. The radical of I is itself an ideal (see Exercise 6 below).

Theorem 4.2, this means that J is the unit ideal. Hence, 1 is an element of J, and so we can write 1 as a linear combination of the polynomial $1 - fX_{m+1}$ and elements of I (with coefficients in $\mathbf{R}' = \mathbb{F}[X_1, \ldots, X_m, X_{m+1}]$). We then make the substitution $X_{m+1} = 1/f$. (More precisely, by taking X_{m+1} to $1/f$ we map $\mathbb{F}[X_1, \ldots, X_m, X_{m+1}]$ to the subring $\mathbb{F}[X][1/f]$ of the field $\mathbb{F}(X)$ of all rational functions of $X_1, \ldots, X_m$.) After we do that, the expression for 1 in terms of the polynomial $1 - fX_{m+1}$ and the elements of I becomes an expression for 1 in terms of elements of I involving just the variables $X_1, \ldots, X_m$ but having f to various powers in the denominator. Multiplying through by f^n for some integer n, we clear denominators. The result is an expression for f^n as a linear combination of elements of I. This proves the theorem. $\square$

Exercises for § 4

1. Describe the maximal ideals of (a) the polynomial ring $\mathbb{F}[X]$ in one variable, where $\mathbb{F}$ is a field (not necessarily algebraically closed); and (b) the polynomial ring $\mathbb{Z}[X]$ in one variable over the ring of integers.

2. Give an example of a sequence of prime ideals $(0) \subset P_1 \subset \cdots \subset P_d$ (where each inclusion is a proper inclusion, and the value of d is given below) in each of the following polynomial rings: (a) $\mathbb{F}[X, Y]$, $d = 2$; (b) $\mathbb{Z}[X]$, $d = 2$; (c) $\mathbb{F}[X_1, \ldots, X_m]$, $d = m$. The maximum value of d for which such a sequence of prime ideals can be found is called the *dimension* of the ring. One can show that the dimension of the ring in part (c) is m – note the agreement with the vector-space dimension of the corresponding space $\mathbb{F}^m$ on which the polynomials are evaluated. Show that an integral domain that is not a field has dimension 1 if and only if every nonzero prime ideal is maximal.

3. Show that a principal ideal domain that is not a field has dimension 1. The converse is not true, however. For example, in the ring $\mathbf{R} = \mathbb{Z}[\sqrt{10}]$ all nonzero prime ideals are maximal. Show that the ideal I generated by 3 and $\sqrt{10} + 1$ is such a (maximal) prime ideal, but is not principal.

4. Let I be the ideal of $\mathbb{Q}[X, Y]$ consisting of all polynomials that vanish at all points of the form $(x, 0)$ and also at the point $(0, 1)$. Find generators for I.

5. Show that the polynomial ring in infinitely many variables over a field (that is, $\bigcup_{m=1}^{\infty} \mathbb{F}[X_1, \ldots, X_m]$) is not Noetherian.

6. Prove that the radical of an ideal is an ideal.

§ 5. Gröbner Bases

An ideal I in the polynomial ring $\mathbf{R} = \mathbb{F}[X] = \mathbb{F}[X_1, \ldots, X_m]$ is usually given to us in the form of a list of generating elements: $I = (f_1, \ldots, f_l)$. Such a set of generating elements is sometimes called a *basis*, even though the representation of an element of I as an $\mathbf{R}$-linear combination of the f_i is certainly not unique.

There are an unlimited number of possible bases for an ideal I, and we would like to find a particularly convenient one. If we had a way of finding a "best possible" basis, then we would be in a much better position to answer various questions about ideals, such as: (1) Given $I = (f_1, \ldots, f_l)$ and $I' = (f'_1, \ldots, f'_{l'})$, are they the same? (2) Given $I = (f_1, \ldots, f_l)$ and an element $f \in \mathbf{R}$, does f belong to I? And if so, how can we express f as an $\mathbf{R}$-linear combination of the f_i? Our goal in this section is to find a way to compute a "particularly convenient" basis for an ideal $I \subset \mathbb{F}[X_1, \ldots, X_m]$.

5.1 Order of Terms

By a "power product" in a polynomial we mean a monomial $X^{\mathbf{i}} = X_1^{i_1} \cdots X_m^{i_m}$ that occurs with nonzero coefficient; and by a "term" of a polynomial we mean a power product taken together with its coefficient, i.e., $a_{\mathbf{i}} X^{\mathbf{i}} = a_{i_1, \ldots, i_m} X_1^{i_1} \cdots X_m^{i_m}$. When looking for an efficient description of a polynomial ideal I, our first item of business is to decide how to order the terms from "highest" to "lowest" in a polynomial such as

$$f(X, Y, Z) = X^3 - X^2Y^2Z + X^2YZ^2 - X^2Z^4 + XY^2 - XZ^3$$
$$+ Y^3Z^3 + Y^2Z + Z^4 \in \mathbb{F}[X, Y, Z] \ ,$$

where $\mathbb{F}$ is some field. The two most common ways are as follows:

1) In the *lexicographical ordering* the terms are listed in the same order in which the power products would appear in a dictionary if they were ordinary words in an alphabet consisting of $X_1, \ldots, X_m$ (or X, Y, Z in the case $m = 3$). The above polynomial is listed in lexicographical order.

2) In the *degree-lexicographical ordering* the power products are listed from highest to lowest total degree, and the terms with a fixed total degree are listed in lexicographical order. For the polynomial $f(X, Y, Z)$ given above, the degree-lexicographical order is

$$f(X, Y, Z) = - X^2Z^4 + Y^3Z^3 - X^2Y^2Z + X^2YZ^2$$
$$- XZ^3 + Z^4 + X^3 + XY^2 + Y^2Z \ . \tag{1}$$

For the rest of this section we shall use the degree-lexicographical ordering unless explicitly stated otherwise.

Many other schemes for ordering the terms are possible. For detailed information on this and other topics discussed in this section we highly recommend [Adams and Loustaunau 1994]; we shall follow the notation and terminology of that textbook. Another readable textbook on the subject is [Cox, Little, and O'Shea 1997].

Notice that if we have any two different power products $X^{\mathbf{i}}$ and $X^{\mathbf{j}}$, either $X^{\mathbf{i}} > X^{\mathbf{j}}$ or else $X^{\mathbf{j}} > X^{\mathbf{i}}$ in the degree-lexicographical ordering. Another important observation is that, given any power product $X^{\mathbf{i}}$, there are only finitely many power products $X^{\mathbf{j}}$ such that $X^{\mathbf{i}} > X^{\mathbf{j}}$.

Definition 5.1. The *leading term* of a polynomial is the first term that appears when the polynomial is listed according to the agreed upon ordering. If $f \in \mathbb{F}[X_1, \ldots, X_m]$, we let $\mathrm{lt}(f)$ denote the leading term of f.

For example, the leading term in the above polynomial $f(X, Y, Z)$ is X^3 in the lexicographical ordering and is $-X^2 Z^4$ in the degree-lexicographical ordering.

5.2 Polynomial Division

Suppose that the leading term of g divides the leading term of f, where $f, g \in \mathbb{F}[X_1, \ldots, X_m]$; in other words, every X_i that appears in $\mathrm{lt}(g)$ appears to at least as great a power in $\mathrm{lt}(f)$. In that case we can get rid of the leading term of f by subtracting a suitable multiple of g – the multiple is the ratio of $\mathrm{lt}(f)$ to $\mathrm{lt}(g)$. More generally, any term of f that is divisible by $\mathrm{lt}(g)$ can be replaced by smaller terms (in the sense of the degree-lexicographical ordering) if we subtract a suitable multiple of g. That gives us the following definition.

Definition 5.2. We say that f *reduces to* h *modulo* g *in one step* if $a_i X^i$ is a term of f that is divisible by $\mathrm{lt}(g)$ and

$$h = f - \frac{a_i X^i}{\mathrm{lt}(g)} g \ .$$

In that case we write

$$f \xrightarrow{\ g\ } h \ .$$

In the important special case when $\mathrm{lt}(f)$ is divisible by $\mathrm{lt}(g)$, we have

$$h = f - \frac{\mathrm{lt}(f)}{\mathrm{lt}(g)} g \ ,$$

and $\mathrm{lt}(h)$ is strictly less than $\mathrm{lt}(f)$ (in the degree-lexicographical ordering).

Example 5.1. Let $f(X, Y, Z) \in \mathbb{F}[X, Y, Z]$ be the polynomial in (1), where $\mathbb{F}$ is any field; and let $g_1(X, Y, Z) = X Z^3 - Y^2 Z^2$. Then f reduces to

$$\begin{aligned}
h_1(X, Y, Z) = {}&- XY^2 Z^3 + Y^3 Z^3 - X^2 Y^2 Z + X^2 Y Z^2 \\
&- X Z^3 + Z^4 + X^3 + XY^2 + Y^2 Z
\end{aligned} \tag{2}$$

modulo g_1 in one step. We can continue the process, since $\mathrm{lt}(g_1)$ divides $\mathrm{lt}(h_1)$. We see that h_1 reduces to

$$\begin{aligned}
h_2(X, Y, Z) = {}&- Y^4 Z^2 + Y^3 Z^3 - X^2 Y^2 Z + X^2 Y Z^2 \\
&- X Z^3 + Z^4 + X^3 + XY^2 + Y^2 Z
\end{aligned} \tag{3}$$

modulo g_1. That is, f reduces to h_2 modulo g_1 in *two* steps. We cannot further reduce the leading term by subtracting multiples of g_1, because $\mathrm{lt}(h_2)$ is not divisible by $\mathrm{lt}(g_1)$. However, if we want, we can make one further reduction, replacing h_2 by $h_2 + g_1 = -Y^4 Z^2 + Y^3 Z^3 - X^2 Y^2 Z + X^2 Y Z^2 - Y^2 Z^2 + Z^4 + X^3 + XY^2 + Y^2 Z$.

Definition 5.3. Let $F = \{g_1, \ldots, g_l\} \subset \mathbb{F}[X_1, \ldots, X_m]$, and let $f \in \mathbb{F}[X_1, \ldots, X_m]$. We say that f *reduces to h modulo the set of polynomials F* if we have a sequence of polynomials beginning with $h_0 = f$ and ending with $h_k = h$ such that h_j reduces to h_{j+1} modulo some $g \in F$ in one step, $j = 0, 1, \ldots, k - 1$.

Example 5.2. Let $F = \{g_1, g_2\}$, where $g_1 = XZ^3 - Y^2Z^2$ and $g_2 = Y^2Z - YZ^2$; and let f be the polynomial in (1). Continuing with Example 5.1, we see that h_2 reduces modulo g_2 to

$$h_3(X, Y, Z) = -X^2Y^2Z + X^2YZ^2 - XZ^3 + Z^4 + X^3 + XY^2 + Y^2Z ,$$

and h_3 reduces modulo g_2 to

$$h_4(X, Y, Z) = -XZ^3 + Z^4 + X^3 + XY^2 + Y^2Z .$$

Next, h_4 reduces modulo g_1 to

$$h_5(X, Y, Z) = -Y^2Z^2 + Z^4 + X^3 + XY^2 + Y^2Z ,$$

and h_5 reduces modulo g_2 to

$$h_6(X, Y, Z) = -YZ^3 + Z^4 + X^3 + XY^2 + Y^2Z .$$

We cannot further lower the leading term, because $\mathrm{lt}(h_6)$ is not divisible by either $\mathrm{lt}(g_1)$ or $\mathrm{lt}(g_2)$. We can perform one more reduction step, because $\mathrm{lt}(g_2)$ divides the last term of h_6; this gives us

$$h(X, Y, Z) = h_7(X, Y, Z) = -YZ^3 + Z^4 + X^3 + XY^2 + YZ^2 .$$

We say that f reduces to $h = h_7$ modulo F, because

$$f \xrightarrow{g_1} h_1 \xrightarrow{g_1} h_2 \xrightarrow{g_2} h_3 \xrightarrow{g_2} h_4 \xrightarrow{g_1} h_5 \xrightarrow{g_2} h_6 \xrightarrow{g_2} h .$$

In Example 5.2, we would have arrived at the same polynomial h if we had reduced h_1 modulo g_2 (rather than g_1), and then reduced the result (which we denote h_2') modulo g_1. That is, we would have obtained the following sequence of reduction steps:

$$f \xrightarrow{g_1} h_1 \xrightarrow{g_2} h_2' \xrightarrow{g_1} h_3 \xrightarrow{g_2} h_4 \xrightarrow{g_1} h_5 \xrightarrow{g_2} h_6 \xrightarrow{g_2} h .$$

However, sometimes when we reduce $f \in \mathbb{F}[X]$ modulo $F = \{g_1, \ldots, g_l\}$ it makes a big difference in what order we choose to take the g_i.

Example 5.3. Let $f(X, Y, Z) = X^2Y^2 + XY \in \mathbb{F}[X, Y, Z]$, where $\mathbb{F}$ is any field; and let $F = \{g_1, g_2, g_3\}$, where $g_1(X, Y, Z) = Y^2 + Z^2$, $g_2(X, Y, Z) = X^2Y + YZ$, and $g_3(X, Y, Z) = Z^3 + XY$. If we first reduce f modulo g_1 we obtain $-X^2Z^2 + XY$, which cannot be reduced further, because neither of its terms is divisible by $\mathrm{lt}(g_1)$, $\mathrm{lt}(g_2)$, or $\mathrm{lt}(g_3)$. On the other hand, if we first reduce f modulo g_2 we obtain $-Y^2Z + XY$, which can be reduced modulo g_1 to get $Z^3 + XY$. Finally, we can reduce the last result modulo g_3 to get 0.

5.3 Gröbner Bases

In Example 5.3 we have an unpleasant situation. If we are lucky enough to start the sequence of reductions by dividing by g_2 (rather than by g_1), then we end up with 0. This means that we can write f as a linear combination of the g_i – namely, $f = Y g_2 - Z g_1 + g_3$. But if we start out by reducing modulo g_1, we get stuck after the first step. For this reason, the set F in Example 5.3 is not a good choice of basis polynomials for the ideal generated by g_1, g_2, g_3. The next definition, which gives one of the most basic notions in computational algebra, provides us with a criterion for making a better choice of basis polynomials.

Definition 5.4. Let $F = \{g_1, \ldots, g_l\} \subset \mathbb{F}[X] = \mathbb{F}[X_1, \ldots, X_m]$ be a finite set of polynomials in m variables over a field $\mathbb{F}$; and let I be the ideal of $\mathbb{F}[X]$ that they generate. We say that F is a *Gröbner basis* for the ideal I if every nonzero $f \in I$ has leading term that is divisible by the leading term of at least one of the g_i.

Theorem 5.1. *In the notation of Definition 5.4, F is a Gröbner basis for I if and only if every $f \in I$ reduces to 0 modulo F.*

Proof. Suppose that F is a Gröbner basis and $f \in I$. Since $\operatorname{lt}(f)$ is divisible by $\operatorname{lt}(g_{i_1})$ for some i_1, we can reduce modulo g_{i_1} to get h_1. Clearly $h_1 \in I$, and h_1 has lower leading term than f in the sense of our chosen order of terms (most likely the degree-lexicographical ordering). We can then repeat the process, reducing h_1 modulo g_{i_2} to get h_2. Since the power product in the leading term gets lower each time, we eventually have to end up with 0.

Conversely, if F is not a Gröbner basis, then there exists $f \in I$ such that $\operatorname{lt}(f)$ is not divisible by $\operatorname{lt}(g_i)$ for any i. Such an f cannot be reduced modulo F to an h with lower leading term. This completes the proof. $\square$

In other words, if we have a Gröbner basis for I, then we have a simple procedure – successive reduction modulo the basis polynomials – for expressing any element of I in terms of the basis. Moreover, given an arbitrary element $f \in \mathbb{F}[X]$, we can use the same method to determine whether or not $f \in I$. Namely, we keep reducing modulo the Gröbner basis until we can go no further. If we have reduced f to 0, then $f \in I$ (and we can explicitly write f in terms of the basis elements); otherwise, $f \notin I$.

In addition, given another ideal I', we can determine whether or not $I' \subset I$: $I' \subset I$ if and only if each of the given generating polynomials of I' is in I. Finally, if we have Gröbner bases for *both* of the ideals I and I', we can easily determine whether or not $I = I'$: equality holds if and only if each element in one basis can be reduced to 0 modulo the polynomials in the other basis.

We shall soon see – in Theorem 5.3 – that every ideal has a Gröbner basis. In order to use Gröbner bases, we need an efficient way to determine if a given basis F is a Gröbner basis and, if not, we need an algorithm to construct a Gröbner basis from F. The difficulty is that there are infinitely many elements in I, and so the criterion in Definition 5.4 (or the one in Theorem 5.1) cannot be checked for all of those elements one by one. Fortunately, it turns out that we need only worry about a small number of elements of I.

Definition 5.5. The *S-polynomial* of two nonzero polynomials $f, g \in \mathbb{F}[X_1, \ldots, X_m]$ is

$$S(f, g) = \frac{L}{\mathrm{lt}(f)} f - \frac{L}{\mathrm{lt}(g)} g ,$$

where L denotes the least common multiple of the leading terms of f and g, that is, the power product of lowest total degree that is divisible by both $\mathrm{lt}(f)$ and $\mathrm{lt}(g)$.

Example 5.4. In Example 5.2 we find that

$$S(g_1, g_2) = XYZ^4 - Y^4Z^2 .$$

In Example 5.3 we have

$$S(g_1, g_2) = X^2Z^2 - Y^2Z ;$$
$$S(g_1, g_3) = Z^5 - XY^3 ;$$
$$S(g_2, g_3) = -X^3Y^2 + YZ^4 .$$

Theorem 5.2 (Buchberger). *In the notation of Definitions 5.4 and 5.5, F is a Gröbner basis for I if and only if $S(g_i, g_j)$ reduces to zero modulo F for every $g_i, g_j \in F$.*

Proof. Since clearly $S(g_i, g_j) \in I$, the "only if" direction is immediate from Theorem 5.1. Now suppose that the condition in the theorem holds. By Definition 5.4, it suffices to show that for any $f \in I$ we have $\mathrm{lt}(g_i)$ dividing $\mathrm{lt}(f)$ for some i.

Without loss of generality, we may suppose that all of the g_i are monic, since both the hypothesis and the conclusion of the theorem are unaffected if each g_i is replaced by the polynomial obtained by dividing g_i by its leading coefficient.

Since $f \in I$, we can write f in the form $f = \sum_{i=1}^{l} h_i g_i$. Let $X^{\mathbf{r}}$ be the largest power product (with respect to the degree-lexicographical ordering) that one finds in the leading term of any of the $h_i g_i$, $i = 1, \ldots, l$. It could easily happen that $X^{\mathbf{r}}$ is larger (with respect to the degree-lexicographical ordering) than the leading term of f, because it could be canceled when we take the sum of all the $h_i g_i$. Of all possible ways that f can be written in the form $\sum_{i=1}^{l} h_i g_i$, suppose that we have chosen a way such that $X^{\mathbf{r}}$ is minimal.

If $X^{\mathbf{r}}$ is the power product in $\mathrm{lt}(f)$, then $\mathrm{lt}(g_i)$ divides $\mathrm{lt}(f)$ for some i, and we are done. So suppose that the power product in $\mathrm{lt}(f)$ is strictly less than $X^{\mathbf{r}}$. To prove the theorem it suffices to show that there is then a way to write f in the form $\sum_{i=1}^{l} h_i' g_i$ with all terms in $h_i' g_i$ smaller than $X^{\mathbf{r}}$, because that would contradict our assumption about minimality of $X^{\mathbf{r}}$.

Consider the products $h_i g_i$ having the power product $X^{\mathbf{r}}$ in its leading term. Without loss of generality, suppose that the first l' products in the sum $f = \sum_{i=1}^{l} h_i g_i$ are the ones having $X^{\mathbf{r}}$ in the leading term. For $i = 1, \ldots, l'$ let $h_i = c_i X^{\mathbf{r}_i} + \tilde{h}_i$, where $\tilde{h}_i$ consists of lower terms, i.e., the power products in $\tilde{h}_i$ are less than $X^{\mathbf{r}_i}$ in the degree-lexicographical ordering. Note that for $i = 1, \ldots, l'$

we have $X^{\mathbf{r}} = X^{\mathbf{r}_i}\mathrm{lt}(g_i)$. For $i = 1, \ldots, l' - 1$ let $X^{\mathbf{s}_i}$ be the least common multiple of $\mathrm{lt}(g_i)$ and $\mathrm{lt}(g_{i+1})$; and let $X^{\mathbf{t}_i} = X^{\mathbf{r}-\mathbf{s}_i}$. Consider the sum

$$
\begin{aligned}
& c_1 X^{\mathbf{t}_1} S(g_1, g_2) + (c_1 + c_2) X^{\mathbf{t}_2} S(g_2, g_3) \\
& + (c_1 + c_2 + c_3) X^{\mathbf{t}_3} S(g_3, g_4) + \cdots \\
& \cdots + (c_1 + c_2 + \cdots + c_{l'-1}) X^{\mathbf{t}_{l'-1}} S(g_{l'-1}, g_{l'}) .
\end{aligned}
\tag{4}
$$

This sum is equal to

$$
\begin{aligned}
& c_1 \left(\frac{X^{\mathbf{r}}}{\mathrm{lt}(g_1)} g_1 - \frac{X^{\mathbf{r}}}{\mathrm{lt}(g_2)} g_2 \right) + (c_1 + c_2) \left(\frac{X^{\mathbf{r}}}{\mathrm{lt}(g_2)} g_2 - \frac{X^{\mathbf{r}}}{\mathrm{lt}(g_3)} g_3 \right) \\
& + (c_1 + c_2 + c_3) \left(\frac{X^{\mathbf{r}}}{\mathrm{lt}(g_3)} g_3 - \frac{X^{\mathbf{r}}}{\mathrm{lt}(g_4)} g_4 \right) + \cdots \\
& \cdots + (c_1 + c_2 + \cdots + c_{l'-1}) \left(\frac{X^{\mathbf{r}}}{\mathrm{lt}(g_{l'-1})} g_{l'-1} - \frac{X^{\mathbf{r}}}{\mathrm{lt}(g_{l'})} g_{l'} \right) \\
& + (c_1 + c_2 + \cdots + c_{l'-1} + c_{l'}) \frac{X^{\mathbf{r}}}{\mathrm{lt}(g_{l'})} g_{l'} ,
\end{aligned}
\tag{5}
$$

since the last coefficient $c_1 + \cdots + c_{l'}$ is zero (this is because the leading power product in $\sum h_i g_i$ is less than $X^{\mathbf{r}}$). On the one hand, the sum in (5) is equal to

$$
c_1 X^{\mathbf{r}_1} g_1 + c_2 X^{\mathbf{r}_2} g_2 + \cdots + c_{l'} X^{\mathbf{r}_{l'}} g_{l'} .
\tag{6}
$$

On the other hand, the sum in (5) is the same as the sum in (4). By assumption, each S-polynomial in (4) can be reduced to 0 modulo F. Because the leading term in $X^{\mathbf{t}_i} S(g_i, g_{i+1})$ is strictly less than $X^{\mathbf{r}}$, the process of reducing $S(g_i, g_{i+1})$ to zero will lead to an expression for $X^{\mathbf{t}_i} S(g_i, g_{i+1})$ in the form $\sum_{j=1}^{l} h_{ij} g_j$ in which $\mathrm{lt}(h_{ij} g_j) < X^{\mathbf{r}}$ for all i, j. Hence, the sum in (6) can be expressed in the form $\sum_{j=1}^{l} h_j'' g_j$, where $\mathrm{lt}(h_j'' g_j) < X^{\mathbf{r}}$ for all j. Then our original polynomial f can be expressed in the form

$$
\begin{aligned}
f = \sum_{i=1}^{l} h_i g_i & = \sum_{i=1}^{l'} (c_i X^{\mathbf{r}_i} + \tilde{h}_i) g_i + \sum_{i=l'+1}^{l} h_i g_i \\
& = \sum_{i=1}^{l'} (h_i'' + \tilde{h}_i) g_i + \sum_{i=l'+1}^{l} (h_i'' + h_i) g_i = \sum_{i=1}^{l} h_i' g_i ,
\end{aligned}
$$

where $h_i' = h_i'' + \tilde{h}_i$ for $i = 1, \ldots, l'$ and $h_i' = h_i'' + h_i$ for $i = l' + 1, \ldots, l$. By construction, all of the power products in $h_i' g_i$ are less than $X^{\mathbf{r}}$. This completes the proof. $\square$

Example 5.5. In Example 5.2, we find that $S(g_1, g_2) = XYZ^4 - Y^4 Z^2$ reduces to $-Y^4 Z^2 + Y^3 Z^3$ modulo g_1 in one step, and $-Y^4 Z^2 + Y^3 Z^3$ reduces to 0 modulo g_2 in one step. Hence F is a Gröbner basis. In Example 5.3, on the other hand, $S(g_1, g_2)$ cannot be reduced to 0 modulo F; hence, F is not a Gröbner basis.

Theorem 5.3. *Let $I \subset \mathbb{F}[X_1, \ldots, X_m]$ be the ideal generated by $F' = \{g_1, \ldots, g_{l'}\}$. Suppose that for any $1 \leq i < j \leq l'$ one reduces the S-polynomial $S(g_i, g_j)$ (see Definition 5.5) modulo F' until a polynomial h_{ij} is obtained that either is 0 or else has leading term that cannot be reduced. In the latter case, h_{ij} is added to the set F'. One continues in this manner, adding $g_{l'+1}, g_{l'+2}, \ldots$ to the set F', until one has a set $F = \{g_1, \ldots, g_l\}$ such that $S(g_i, g_j)$ reduces to 0 modulo F for all $1 \leq i < j \leq l$. This algorithm terminates in a finite number of steps, and gives a Gröbner basis of I.*

Proof. For each j with $l' \leq j \leq l$ let J_j be the ideal generated by $\mathrm{lt}(g_1)$, $\mathrm{lt}(g_2)$, $\ldots$, $\mathrm{lt}(g_j)$. By construction, each $\mathrm{lt}(g_j)$ for $j > l'$ is not divisible by any of the earlier $\mathrm{lt}(g_1)$, $\mathrm{lt}(g_2)$, $\ldots$, $\mathrm{lt}(g_{j-1})$. Thus, the ideals

$$J_{l'} \subset J_{l'+1} \subset J_{l'+2} \subset \cdots$$

form a strictly increasing sequence. By Corollary 4.2 of the Hilbert Basis Theorem, there can be only finitely many ideals, and hence only finitely many g_j. Thus, the algorithm terminates. The resulting set F is a Gröbner basis by Theorem 5.2. □

Example 5.6. In Example 5.5 we saw that the set $F' = \{g_1, g_2, g_3\}$, where $g_1 = Y^2 + Z^2$, $g_2 = X^2Y + YZ$, $g_3 = Z^3 + XY$, is not a Gröbner basis for the ideal I it generates, because $S(g_1, g_2) = X^2Z^2 - Y^2Z$ cannot be reduced to 0. If we set $g_4 = S(g_1, g_2)$ and $F = \{g_1, g_2, g_3, g_4\}$, we easily check that all of the following polynomials reduce to 0 modulo F: $S(g_1, g_2) = g_4$, $S(g_1, g_3) = Z^5 - XY^3$, $S(g_1, g_4) = X^2Z^4 + Y^4Z$, $S(g_2, g_3) = -X^3Y^2 + YZ^4$, $S(g_2, g_4) = Y^3Z + YZ^3$, and $S(g_3, g_4) = X^3Y + Y^2Z^2$. Thus, F is a Gröbner basis for I.

5.4 Reduced Gröbner Bases

Definition 5.6. A Gröbner basis $\{g_1, \ldots, g_l\}$ of an ideal $I \subset \mathbb{F}[X_1, \ldots, X_m]$ is said to be *minimal* if all of the g_i are monic and if $\mathrm{lt}(g_i)$ does not divide $\mathrm{lt}(g_j)$ for $i \neq j$, $i, j = 1, \ldots, l$.

Given a Gröbner basis F, all we need to do to obtain a minimal Gröbner basis is to successively remove from F any g_i whose leading term is divisible by the leading term of another element of F. More precisely, suppose that we have a Gröbner basis of I consisting of $l + 1$ polynomials $g_1, \ldots, g_{l+1}$, one of whose leading term divides the leading term of another. Without loss of generality, we suppose that $\mathrm{lt}(g_1)$ divides $\mathrm{lt}(g_{l+1})$. Suppose that g_{l+1} reduces to h modulo g_1 in one step, where $\mathrm{lt}(h) < \mathrm{lt}(g_{l+1})$. Since $h \in I$, h can be reduced to 0 modulo the set $\{g_1, \ldots, g_l, g_{l+1}\}$. However, g_{l+1} cannot be used in reducing h to 0, because its leading term is greater than $\mathrm{lt}(h)$; hence h is a linear combination of $g_1, \ldots, g_l$. Then g_{l+1} is also a linear combination of $g_1, \ldots, g_l$, and so $\{g_1, \ldots, g_l\}$ is a basis for I. It is a Gröbner basis because, if the leading term of $f \in I$ is divisible by $\mathrm{lt}(g_{l+1})$, it is also divisible by $\mathrm{lt}(g_1)$. Thus, the criterion in Definition 5.4 still holds after g_{l+1} has been removed from the set.

Definition 5.7. A Gröbner basis $\{g_1, \ldots, g_l\}$ of an ideal $I \subset \mathbb{F}[X_1, \ldots, X_m]$ is said to be *reduced* if all of the g_i are monic and if none of the terms of g_i is divisible by $\mathrm{lt}(g_j)$ for $j \neq i$.

Example 5.7. The Gröbner basis in Example 5.6 is minimal, but it is not reduced, because the term $-Y^2 Z$ in g_4 is divisible by $\mathrm{lt}(g_1)$. If we reduce g_4 modulo g_1 and then modulo g_3, and if we replace g_4 by $g_4' = g_4 + Zg_1 - g_3 = X^2 Z^2 - XY$, we obtain the reduced Gröbner basis $\{g_1, g_2, g_3, g_4'\}$.

Once we have a minimal Gröbner basis $\{g_1, \ldots, g_l\}$ we can obtain a reduced Gröbner basis as follows. First reduce g_1 modulo $g_2, \ldots, g_l$ until no term of the resulting polynomial h_1 is divisible by $\mathrm{lt}(g_i)$ for $i = 2, \ldots, l$. Then replace g_1 by h_1. Next, reduce g_2 modulo $h_1, g_3, \ldots, g_l$ until no term of the resulting polynomial h_2 is divisible by $\mathrm{lt}(h_1)$ or $\mathrm{lt}(g_i)$ for $i = 3, \ldots, l$; and replace g_2 by h_2. Continue in this manner until $g_1, \ldots, g_l$ have been replaced by $h_1, \ldots, h_l$. It is easy to see that $\{h_1, \ldots, h_l\}$ is a reduced Gröbner basis.

In summary, given a set of generators F for an ideal $I \subset \mathbb{F}[X_1, \ldots, X_m]$, here is how to obtain a reduced Gröbner basis:

1) For each pair of polynomials in F, compute the S-polynomial and reduce it modulo F. After reducing it as far as possible, add the reduced polynomial to F if it is not 0. Continue until all S-polynomials of pairs of elements of the expanded F reduce to 0.
2) Go through the polynomials in F, deleting the ones whose leading term is divisible by the leading term of another polynomial in the list.
3) Make the polynomials in F monic by dividing each one by its leading coefficient.
4) Successively replace each polynomial in F by the result obtained by reducing it modulo all of the other elements of F.

Theorem 5.4. *Every ideal of* $\mathbb{F}[X_1, \ldots, X_m]$ *has a unique reduced Gröbner basis.*

Proof. The above procedure gives a reduced Gröbner basis. It remains to prove uniqueness. Suppose that $\{g_1, \ldots, g_l\}$ and $\{h_1, \ldots, h_{l'}\}$ are two reduced Gröbner bases for the same ideal. By Exercise 3 below, $l = l'$ and we may assume that $\mathrm{lt}(g_i) = \mathrm{lt}(h_i)$ for $i = 1, \ldots, l$. We claim that $g_i = h_i$ for $i = 1, \ldots, l$. Otherwise, since $g_i - h_i \in I$, by Definition 5.4 the leading term of $g_i - h_i$ would have to be divisible by $\mathrm{lt}(g_j) = \mathrm{lt}(h_j)$ for some j. Clearly $j \neq i$, since $\mathrm{lt}(g_i - h_i) < \mathrm{lt}(g_i)$. Then a term of either g_i or h_i would be divisible by $\mathrm{lt}(g_j) = \mathrm{lt}(h_j)$, contradicting Definition 5.7. $\square$

Exercises for § 5

Use the degree-lexicographical ordering in these exercises.

1. Suppose that $F = \{g_1, \ldots, g_l\} \subset \mathbb{F}[X_1, \ldots, X_m]$ is a set of linear forms (polynomials of total degree 1 with zero constant term). Let I be the ideal generated by

F. Give necessary and sufficient conditions for F to be: (a) a minimal Gröbner basis; (b) a reduced Gröbner basis.

2. Suppose that $F = \{g_1, \ldots, g_l\} \subset \mathbb{F}[X]$ is a set of polynomials in one variable. Let I be the ideal generated by F. Give necessary and sufficient conditions for F to be a reduced Gröbner basis.

3. Prove that if $F = \{g_1, \ldots, g_l\}$ and $F' = \{g_1', \ldots, g_{l'}'\}$ are both minimal Gröbner bases of the same ideal I, then $l = l'$. Also prove that (after renumbering if necessary) one has $\text{lt}(g_i)=\text{lt}(g_i')$, $i = 1, \ldots, l$.

4. True or False? Please explain.

(a) Generalizing the case when $m = 1$ (see Exercise 2), one can find an upper bound in terms of m for the number of elements in a reduced Gröbner basis of an ideal in $\mathbb{F}[X_1, \ldots, X_m]$.

(b) The number of elements in a basis for an ideal I of $\mathbb{F}[X_1, \ldots, X_m]$ is always greater than or equal to the number of elements in a minimal Gröbner basis.

(c) If G is a Gröbner basis, then the set of all polynomials that cannot be reduced modulo G (see Definition 5.3) is a set of representatives for the quotient ring $\mathbb{F}[X]/I$.

5. Find the reduced Gröbner basis for the ideal consisting of all polynomials in $\mathbb{F}[X, Y]$ that vanish at the two points $(0, 0)$ and $(1, 1)$.

6. Find the reduced Gröbner basis for the ideal in $\mathbb{F}[X_1, \ldots, X_m]$ consisting of all polynomials whose power products are all of total degree at least n.

7. Find the reduced Gröbner basis for the ideal in $\mathbb{F}[X, Y, Z]$ generated by $g_1 = XZ$, $g_2 = XY - Z$, and $g_3 = YZ - X$.

8. Find the reduced Gröbner basis for the ideal in $\mathbb{F}[X, Y]$ generated by $g_1 = X^2Y - Y$, $g_2 = Y^2 - X$, and $g_3 = X^2Y^2 - XY$.

9. Find the reduced Gröbner basis for the ideal in $\mathbb{F}[X, Y, Z]$ generated by $g_1 = X^3 - YZ$, $g_2 = Y^3 - XZ$, and $g_3 = XY - Z$.

10. Find the reduced Gröbner basis for the ideal in $\mathbb{F}[X, Y]$ generated by $g_1 = X^2 - Y^2$, $g_2 = X^3 - Y^3$, and $g_3 = X^2Y - XY^2$.

11. Find the reduced Gröbner basis for the ideal in $\mathbb{F}[X, Y, Z]$ generated by $g_1 = X^3 - Y$, $g_2 = Y^3 - X$, and $g_3 = X^2Y^2 - XY$.

12. Suppose that I is an ideal of $\mathbb{F}[X] = \mathbb{F}[X_1, \ldots, X_m]$. Let $\overline{\mathbb{F}}$ denote the algebraic closure of $\mathbb{F}$.

(a) Suppose that $f \in \mathbb{F}[X]$ can be written as a linear combination of elements of I with coefficients in $\overline{\mathbb{F}}[X]$. Prove that f can be written as a linear combination of elements of I with coefficients in $\mathbb{F}[X]$.

(b) Let G be a Gröbner basis for I. Let $\overline{I}$ be the ideal of $\overline{\mathbb{F}}[X]$ generated by the elements of I. Show that G is a Gröbner basis for $\overline{I}$.

13. Let G be a Gröbner basis for an ideal I in $\mathbb{F}[X_1, \ldots, X_m]$. Suppose that there are only finitely many points (with coordinates in the algebraic closure of $\mathbb{F}$) where all of the polynomials in I vanish. Prove that for each $i = 1, \ldots, m$ there exists an element of G whose leading term is of the form cX_i^l.

Chapter 4. Hidden Monomial Cryptosystems

§ 1. The Imai–Matsumoto System

1.1 The System

Let $\mathbb{K}$ be an extension of degree n of the finite field $\mathbb{F}_q$, where q is a power of 2, and let $\beta_1, \beta_2, \ldots, \beta_n \in \mathbb{K}$ be a basis of $\mathbb{K}$ as an $\mathbb{F}_q$-vector space. Alice will be using the Imai–Matsumoto system in $\mathbb{K}$. She regards each element of $\mathbb{K}$ as an n-tuple over $\mathbb{F}_q$. Alice may choose to keep her basis secret, in which case we cannot assume that a cryptanalyst (whom we shall name "Catherine") knows what basis she is using.

Both plaintext message units and ciphertext message units will be n-tuples over $\mathbb{F}_q$. We will use the vector notation $\overline{x} = (x_1, \ldots, x_n) \in \mathbb{F}_q^n$ for plaintext and $\overline{y} = (y_1, \ldots, y_n) \in \mathbb{F}_q^n$ for ciphertext. When working with matrices, we shall consider vectors to be column-vectors (although in the text we shall continue writing them as rows).

In transforming plaintext into ciphertext, Alice will work with two intermediate vectors, denoted $\overline{u} = (u_1, \ldots, u_n) \in \mathbb{F}_q^n$ and $\overline{v} = (v_1, \ldots, v_n) \in \mathbb{F}_q^n$. Given a vector in $\mathbb{F}_q^n$, we shall use boldface to denote the corresponding element of $\mathbb{K}$ with respect to the basis β_j. For example, if $\overline{u} = (u_1, \ldots, u_n) \in \mathbb{F}_q^n$, then we set $\mathbf{u} = u_1\beta_1 + \cdots + u_n\beta_n \in \mathbb{K}$.

Next, Alice chooses an exponent h, $0 < h < q^n$, that is of the form

$$h = q^\theta + 1$$

and satisfies the condition g.c.d.$(h, q^n - 1) = 1$. (Recall that q was chosen to be a power of 2; if q were odd, then g.c.d.$(h, q^n - 1)$ would be at least 2.) The condition g.c.d.$(h, q^n - 1) = 1$ is equivalent to requiring that the map $\mathbf{u} \mapsto \mathbf{u}^h$ on $\mathbb{K}$ is one-to-one; its inverse is the map $\mathbf{u} \mapsto \mathbf{u}^{h'}$, where h' is the multiplicative inverse of h modulo $q^n - 1$.

Alice may choose to keep h secret. However, since there are relatively few possible values for h, she must assume that Catherine will be prepared to run through all possibilities for h. That is, even if she keeps h secret, the security of her system must lie elsewhere.

In addition, Alice chooses two secret affine transformations, i.e., two invertible $n \times n$-matrices $A = \{a_{ij}\}_{1 \le i,j \le n}$ and $B = \{b_{ij}\}_{1 \le i,j \le n}$ with entries in $\mathbb{F}_q$, and two constant vectors $\overline{c} = (c_1, \ldots, c_n)$ and $\overline{d} = (d_1, \ldots, d_n)$. The purpose of the two affine transformations is to "hide the monomial map" $\mathbf{u} \mapsto \mathbf{u}^h$ – hence the name "hidden monomial cryptosystem".

We now describe how Alice gets her public rule for going from plaintext $\overline{x} \in \mathbb{F}_q^n$ to ciphertext $\overline{y} \in \mathbb{F}_q^n$. First, she sets

$$\overline{u} = A\overline{x} + \overline{c} .$$

Next, she would like to have $\mathbf{v} \in \mathbb{K}$ simply equal to the h-th power of $\mathbf{u} \in \mathbb{K}$, and then set

$$\overline{y} = B^{-1}(\overline{v} - \overline{d}) , \qquad (\text{that is, } \overline{v} = B\overline{y} + \overline{d}) ,$$

where $\overline{v} \in \mathbb{F}_q^n$ is the vector corresponding to $\mathbf{v} \in \mathbb{K}$. However, her public encryption rule will go right from $\overline{x}$ to $\overline{y}$, and will not directly involve exponentiation at all.

In order to get formulas going from $\overline{x}$ directly to $\overline{y}$, Alice notices that, since $\mathbf{v} = \mathbf{u}^h$ and $h = q^\theta + 1$, she has

$$\mathbf{v} = \mathbf{u}^{q^\theta} \cdot \mathbf{u} . \tag{1}$$

Recall that for any $k = 1, 2, \ldots, n$ the operation of raising to the q^k-th power in $\mathbb{K}$ is an $\mathbb{F}_q$-linear transformation. Let $P^{(k)} = \{p_{ij}^{(k)}\}_{1 \le i, j \le n}$ be the matrix of this linear transformation in the basis $\beta_1, \ldots, \beta_n$, i.e.,

$$\beta_i^{q^k} = \sum_{j=1}^n p_{ij}^{(k)} \beta_j , \qquad p_{ij}^{(k)} \in \mathbb{F}_q , \tag{2}$$

for $1 \le i, k \le n$. Alice also writes all products of basis elements in terms of the basis, i.e.,

$$\beta_i \beta_j = \sum_{l=1}^n m_{ijl} \beta_l , \qquad m_{ijl} \in \mathbb{F}_q , \tag{3}$$

for each $1 \le i, j \le n$. Now equation (1) can be expanded to give

$$\sum_{1 \le l \le n} v_l \beta_l = \left(\sum_{i=1}^n u_i \beta_i^{q^\theta} \right) \left(\sum_{j=1}^n u_j \beta_j \right)$$

$$= \left(\sum_{1 \le i, \mu \le n} p_{i\mu}^{(\theta)} u_i \beta_\mu \right) \left(\sum_{j=1}^n u_j \beta_j \right) . \tag{4}$$

If we use (3) and then compare the coefficients of β_l on the left and right sides of (4), for each l we obtain

$$v_l = \sum_{1 \le i, j, \mu \le n} p_{i\mu}^{(\theta)} m_{\mu j l} u_i u_j . \tag{5}$$

Of course, Alice knows all of the coefficients $m_{\mu j l}$ and $p_{i\mu}^{(\theta)}$. She now uses her affine relations

$$\overline{u} = A\overline{x} + \overline{c} , \qquad \overline{v} = B\overline{y} + \overline{d} , \tag{6}$$

to replace u_i by $c_i + \sum_\rho a_{i\rho} x_\rho$ and replace v_l by $d_l + \sum_\sigma b_{l\sigma} y_\sigma$ in (5). When she gathers coefficients of each product $x_i x_j$, she obtains n equations, each with a polynomial of total degree 2 in the $x_1, \ldots, x_n$ on the right and with a linear expression in the $y_1, \ldots, y_n$ on the left. Using linear algebra, she can get n equations that express each y_i as a polynomial of total degree 2 in the $x_1, \ldots, x_n$.

Alice makes these n equations public. If Bob wants to send her a plaintext message $\overline{x}$, he substitutes the x_i in these equations and finds the y_j. On the other hand, Catherine, who knows only the ciphertext (and the public key), must solve a *nonlinear* system for the unknowns x_i.

When Alice receives the ciphertext $\overline{y}$, she uses her knowledge of A, B, $\overline{c}$, $\overline{d}$, and h to recover $\overline{x}$, without having to solve the publicly known equations for the x_i. Namely, let h' be the multiplicative inverse of h modulo $q^n - 1$, so that the map $\mathbf{u} = \mathbf{v}^{h'}$ inverts the map $\mathbf{v} = \mathbf{u}^h$ on $\mathbb{K}$. Alice first computes $\overline{v} = B\overline{y} + \overline{d}$, then raises $\mathbf{v} = \sum v_i \beta_i \in \mathbb{K}$ to the h'-th power (i.e., sets $\mathbf{u} = \mathbf{v}^{h'}$), and finally computes $\overline{x} = A^{-1}(\overline{u} - \overline{c})$.

The following diagram summarizes Alice's decryption:

$$y_1, \ldots, y_n$$
$$\Downarrow$$
$$\overline{v} = B\overline{y} + \overline{d}$$
$$\Downarrow$$
$$\mathbf{v} = \sum v_i \beta_i$$
$$\Downarrow$$
$$\mathbf{u} = \mathbf{v}^{h'}$$
$$\Downarrow$$
$$\overline{x} = A^{-1}(\overline{u} - \overline{c}) \ .$$

Remark. The cryptosystem described above is a simplified version of the one proposed in the original paper [Imai and Matsumoto 1989]. Instead of using a single extension $\mathbb{K}$ of degree n over $\mathbb{F}_q$, they wrote n as a sum $n = n_1 + \cdots + n_d$ and used d extensions $\mathbb{K}_1, \ldots, \mathbb{K}_d$, where $\mathbb{K}_i$ has degree n_i over $\mathbb{F}_q$. They chose a different exponent $h_i = q^{\theta_i} + 1$ in each $\mathbb{K}_i$, where g.c.d.$(h_i, q^{n_i} - 1) = 1$. The n components of the vector u were split up into d subsets of n_i components, and the corresponding element $\mathbf{u}_i \in \mathbb{K}_i$ was transformed to $\mathbf{v}_i \in \mathbb{K}_i$ by raising it to the h_i-th power.

At first, this greater generality might seem to contribute to the security of the system. However, it turns out that the cryptanalysis in §1.2 goes through just as well for this more general system. For details about breaking the original Imai–Matsumoto system, see [Patarin 1995].

Example 1.1. Here is a "toy example" of the Imai–Matsumoto system. That means that its purpose is to illustrate the mechanical operation of the cryptosystem, but its parameters are too small to give any security. Let $q = 2$, $n = 5$, and let $\mathbb{K}$ be represented as the set of polynomials in $\mathbb{F}_2[X]$ modulo the irreducible polynomial $f(X) = X^5 + X^4 + X^3 + X + 1$. We use the basis $\{\beta_1, \beta_2, \beta_3, \beta_4, \beta_5\} = \{1, X, X^2, X^3, X^4\}$. Further let $\theta = 3$, $h = 9$, $h' = 7$, and set

$$A = \begin{pmatrix} 1 & 0 & 1 & 1 & 0 \\ 0 & 1 & 1 & 0 & 1 \\ 1 & 1 & 0 & 0 & 1 \\ 0 & 1 & 0 & 1 & 0 \\ 0 & 0 & 0 & 1 & 1 \end{pmatrix}, \qquad B = \begin{pmatrix} 1 & 0 & 0 & 1 & 1 \\ 0 & 0 & 1 & 1 & 0 \\ 1 & 1 & 0 & 0 & 1 \\ 1 & 1 & 0 & 0 & 0 \\ 1 & 0 & 0 & 0 & 0 \end{pmatrix},$$

$$\overline{c} = (1, 0, 1, 1, 1), \qquad \overline{d} = (1, 0, 1, 0, 0).$$

Then

$$A^{-1} = \begin{pmatrix} 0 & 0 & 1 & 1 & 1 \\ 1 & 1 & 1 & 1 & 0 \\ 0 & 1 & 0 & 1 & 1 \\ 1 & 1 & 1 & 0 & 0 \\ 1 & 1 & 1 & 0 & 1 \end{pmatrix}, \qquad B^{-1} = \begin{pmatrix} 0 & 0 & 0 & 0 & 1 \\ 0 & 0 & 0 & 1 & 1 \\ 1 & 1 & 1 & 1 & 1 \\ 1 & 0 & 1 & 1 & 1 \\ 0 & 0 & 1 & 1 & 0 \end{pmatrix}.$$

Thus, the $\overline{u}$-vector is expressed in terms of the $\overline{x}$-vector as follows: $u_1 = x_1 + x_3 + x_4 + 1$, $u_2 = x_2 + x_3 + x_5$, $u_3 = x_1 + x_2 + x_5 + 1$, $u_4 = x_2 + x_4 + 1$, $u_5 = x_4 + x_5 + 1$. If we write $\mathbf{v} = \mathbf{u}^9 = (u_1 + u_2 X + u_3 X^2 + u_4 X^3 + u_5 X^4)(u_1 + u_2 X^8 + u_3 X^{16} + u_4 X^{24} + u_5 X^{32})$ and reduce the product on the right modulo $f(X)$, we find that $\overline{v}$ is expressed in terms of $\overline{x}$ as follows:

$$\begin{aligned} v_1 &= 1 + x_1^2 + x_1 x_3 + x_1 x_2 + x_4 + x_4 x_5 + x_1 x_4 + x_2 x_4 \\ &\quad + x_1 + x_2 + x_3 x_5 + x_2^2 \\ v_2 &= x_5 x_1 + x_3 x_2 + x_1^2 + x_2 x_5 + x_5^2 + x_4 + x_1 x_4 + x_1 \\ &\quad + x_3^2 + x_2 + x_3 x_5 \\ v_3 &= x_1 x_3 + x_1 + x_1 x_2 + x_3 x_2 + x_3 x_4 + x_2 + x_3 + x_4^2 + x_3 x_5 + x_2^2 \\ v_4 &= x_3 x_4 + x_1^2 + x_5^2 + x_3 + 1 + x_1 x_3 + x_1 x_4 + x_2 x_4 + x_4^2 + x_2^2 \\ v_5 &= x_3 x_2 + 1 + x_5 x_1 + x_3 + x_5 + x_5^2 + x_1 x_3 + x_1 x_2 + x_4 \\ &\quad + x_1 x_4 + x_3^2 + x_2 + x_4^2 + x_3 x_5 . \end{aligned}$$

Finally, the public equations relating $\overline{y}$ to $\overline{x}$ are:

$$\begin{aligned} y_1 &= x_3 x_2 + 1 + x_5 x_1 + x_3 + x_5 + x_5^2 + x_1 x_3 + x_1 x_2 + x_4 + x_1 x_4 \\ &\quad + x_3^2 + x_2 + x_4^2 + x_3 x_5 \\ y_2 &= x_3 x_4 + x_1^2 + x_2 x_4 + x_2^2 + x_3 x_2 + x_5 x_1 + x_5 + x_1 x_2 + x_4 \\ &\quad + x_3^2 + x_2 + x_3 x_5 \\ y_3 &= 1 + x_1^2 + x_1 + x_3 + x_4 + x_5 + x_4^2 + x_1 x_2 + x_4 x_5 + x_3 x_2 \\ &\quad + x_2^2 + x_2 x_5 + x_5^2 \\ y_4 &= 1 + x_1 x_4 + x_3^2 + x_2 + x_3 + x_5 + x_4^2 + x_3 x_5 + x_5 x_1 + x_1 x_2 + x_4 x_5 + x_2^2 \\ y_5 &= x_1 + x_1 x_2 + x_3 x_2 + x_2 + x_3 x_5 + x_1^2 + x_5^2 + x_1 x_4 + x_2 x_4 . \end{aligned}$$

1.2 Patarin's Cryptanalysis

At Crypto '95, Jacques Patarin showed how to break the Imai–Matsumoto cryptosystem. His idea, though ingenious, is actually quite simple. He noticed that if one takes the equation $\mathbf{v} = \mathbf{u}^h = \mathbf{u}^{q^\theta+1}$, raises both sides to the $(q^\theta - 1)$-th power, and multiplies both sides by $\mathbf{uv}$, one gets an equation

$$\mathbf{u} \cdot \mathbf{v}^{q^\theta} = \mathbf{u}^{q^{2\theta}} \cdot \mathbf{v} \tag{7}$$

that leads to equations in $x_1, \ldots, x_n, y_1, \ldots, y_n$ that are linear in *both* sets of variables. Using linear algebra, Catherine the cryptanalyst can find these equations even if she has no idea what Alice's parameters are. These equations probably won't be quite enough to uniquely determine the plaintext from the ciphertext. But they will reduce the search for the plaintext $\overline{x}$ to a small enough affine subspace of $\mathbb{F}_q^n$ so that, in all likelihood, even an exhaustive search will be feasible. We now give more details.

Catherine knows, of course, that Alice is using the Imai–Matsumoto cryptosystem in a field extension $\mathbb{K}$ of degree n over $\mathbb{F}_q$ (where q is a power of 2). She thus knows that an equation of the form (7) holds, and that there are linear relations (2) and (3) and affine relations (6) that together lead to equations of the form

$$\left(\sum_{1 \leq i,j \leq n} \alpha_{ijl} x_i y_j \right) + \left(\cdot \sum_{1 \leq i \leq n} (\beta_{il} x_i + \gamma_{il} y_i) \right) + \delta_l = 0 \;, \tag{8}$$

$l = 1, \ldots, n$. Equation (8) is derived in exactly the same way in which, starting from the right hand side of (1), we obtained polynomials of total degree 2 in $x_1, \ldots, x_n$ (see equations (1) through (6)). To break the Imai–Matsumoto cryptosystem, Catherine's strategy is to ignore the public equations, and instead find the much better equations (8) that are linear in *both* sets of variables $\overline{x}$ and $\overline{y}$.

At first Catherine has no idea what the coefficients in these relations are, because she does not know the coefficients in (6), (2), or (3) (since she does not even know the basis $\beta_1, \ldots, \beta_n$). However, she can generate a large number of plaintext-ciphertext pairs $(x_1, \ldots, x_n, y_1, \ldots, y_n)$ by simply using Alice's public equations (just as Bob would do in order to send Alice messages). Any such $2n$-tuple can be substituted into (8) to yield a linear equation in the $(n+1)^2$ unknown coefficients $\alpha_{ij}, \beta_i, \gamma_i, \delta$ in an equation of the form (8).

In this way Catherine will eventually be able to find a maximal set of L linearly independent equations of the form (8) that are satisfied by all plaintext-ciphertext pairs. We know that many such equations will come from (7), and it is possible (though probably not likely) that there will be some other equations of the form (8) that do not come from (7). To simplify the argument that follows, let us assume that all of the equations (8) that are satisfied by all plaintext-ciphertext pairs actually come from (7). If there are any additional equations, then they will only help Catherine break the cryptosystem more easily. Suppose that there are L independent equations indexed by $l = 1, \ldots, L$, as in (8). Will they be enough to uniquely determine the plaintext $\overline{x}$ corresponding to a given ciphertext $\overline{y}$? Probably not, for the following reason.

When Catherine raised both sides of the equation $\mathbf{v} = \mathbf{u}^h$ to the $(q^\theta - 1)$-th power and then multiplied through by $\mathbf{uv}$, she lost information. To put it another way, new "extraneous" solutions – ones that do not correspond to plaintext-ciphertext pairs – were introduced by exponentiating both sides of the equation. This difficulty is familiar from high school algebra, where one might remove a radical in an equation by squaring both sides, only to find that the resulting solution is extraneous, i.e., not a solution of the original equation. (For example, the equation $\sqrt{x+1} - \sqrt{x} = 2$, which has no real solutions, can be "solved" to get the extraneous solution $x = 9/16$.)

Let us suppose that Catherine has found a complete set of independent equations (8) that are satisfied by all plaintext-ciphertext pairs. She then intercepts a ciphertext vector $\overline{y}_0$ and substitutes its coordinates into all of the equations (8). At that point she has L linear equations in the n unknowns $x_1, \ldots, x_n$. How many of these L equations are independent? In general, this number – let's call it Λ – will be less than L, and it may depend upon the particular $\overline{y}_0$.

Equivalently, how many different solutions $\overline{x}$ will the system (8) have (after $\overline{y}_0$ has been substituted in place of $\overline{y}$)? We know that the plaintext vector provides one solution, and so the equations are consistent. From linear algebra we then know that the space of solutions is an $(n - \Lambda)$-dimensional affine space in $\mathbb{F}_q^n$ (by an "affine" space we mean a linear subspace shifted by a constant vector). In other words, when we regard the equations (8) as a system of linear equations in $x_1, \ldots, x_n$ after the substitution $\overline{y} = \overline{y}_0$, it will have exactly $q^{n-\Lambda}$ solutions.

On the other hand, these equations in $x_1, \ldots, x_n$ are equivalent to the equation (7) (regarded as an equation in the unknown $\mathbf{u}$ after a specific value $\mathbf{v}_0$, where $\overline{v}_0 = B\overline{y}_0 + \overline{d}$, has been substituted for $\mathbf{v}$). Here remember that, for simplicity, we are assuming that all of the equations in (8) come from (7). That is, the solutions of (7) are in one-to-one correspondence with the solutions to the system (8). When $\mathbf{v} = \mathbf{v}_0$ is fixed, how many solutions $\mathbf{u}$ are there to (7)? First, there is the trivial solution $\mathbf{u} = 0$. Then there is the unique solution $\mathbf{u}_0 = \mathbf{v}_0^{h'}$ of the equation $\mathbf{v} = \mathbf{u}^h$ that was raised to the $(q^\theta - 1)$-th power to get (7). If $\mathbf{u}$ is another nonzero solution of (7) with $\mathbf{v} = \mathbf{v}_0$, then we have both

$$\mathbf{v}_0^{q^\theta - 1} = \mathbf{u}_0^{h(q^\theta - 1)} \quad \text{and} \quad \mathbf{v}_0^{q^\theta - 1} = \mathbf{u}^{h(q^\theta - 1)} .$$

Hence, $\mathbf{u}_0^{h(q^\theta - 1)} = \mathbf{u}^{h(q^\theta - 1)}$. Raising both sides of the last equality to the h'-th power (which inverts the h-th power map in $\mathbb{K}$), we find that $\mathbf{u}_0^{q^\theta - 1} = \mathbf{u}^{q^\theta - 1}$. This means that $\mathbf{u}$ differs from $\mathbf{u}_0$ by a factor that is a $(q^\theta - 1)$-th root of unity in $\mathbb{K}$. Conversely, if ζ is any $(q^\theta - 1)$-th root of unity in $\mathbb{K}$, then clearly $\mathbf{u} = \zeta \mathbf{u}_0$ will be a nonzero solution of (7).

How many $(q^\theta - 1)$-th roots of unity are there in $\mathbb{K}$? Since the nonzero elements of $\mathbb{K}$ form a cyclic group of order $q^n - 1$, there are g.c.d.$(q^\theta - 1, q^n - 1)$ such roots. (See Exercise 12 of §2 in Chapter 3.) By Exercise 1 below, g.c.d.$(q^\theta - 1, q^n - 1) = q^d - 1$, where $d = $ g.c.d.(θ, n). When we count the zero solution, we conclude that there are exactly q^d solutions $\mathbf{u}$ of (7) (with $\mathbf{v} = \mathbf{v}_0$), and hence q^d solutions $\bar{x}$ of (8) (with $\bar{y} = \bar{y}_0$). That is,

$$n - \Lambda = d = \text{g.c.d.}(\theta, n) . \tag{9}$$

The number in (9) is a measure of how far Catherine is from uniquely determining $\bar{x}$ from her equations (8). She can use the equations (8) to determine a d-dimensional affine space that contains the desired plaintext vector. Then she has to search among the q^d vectors to find the plaintext.

What is the largest that d can be? Since θ was chosen so that $\text{g.c.d.}(q^\theta + 1, q^n - 1) = 1$, it is easy to rule out $d = n$ and $d = n/2$. It is, however, possible to have $d = \theta = n/3$ (see Exercise 3 below).

We conclude that Catherine has to search through a space that has at most 1/3 the dimension of the entire space of possible plaintexts. This means that the Imai–Matsumoto system is either insecure or inefficient. That is, even if $\theta = n/3$, in order to make the system resistant to exhaustive search attacks, one must choose n to be 3 times larger than originally thought.

Despite the weakness in their system, Imai and Matsumoto contributed a valuable idea for a cryptosystem. Soon after breaking the particular system that they had proposed, Patarin found ways to modify it so as to resist attacks such as the one described above. These modifications will be the subject of §§2–3.

Exercises for § 1

Let a, b, n, θ, and q be positive integers with $q \geq 2$. Prove:

1. $\text{g.c.d.}(q^a - 1, q^b - 1) = q^{\text{g.c.d.}(a,b)} - 1$.

2. If q is even and $\text{g.c.d.}(2\theta, n) = 1$, then $\text{g.c.d.}(q^\theta + 1, q^n - 1) = 1$.

3. If q is even and n is an odd multiple of θ, then $\text{g.c.d.}(q^\theta + 1, q^n - 1) = 1$.

4. In Example 1.1, encrypt the following plaintext vectors using the public equations: (a) $(0, 1, 0, 0, 0)$; (b) $(1, 1, 1, 1, 1)$; (c) $(1, 0, 0, 1, 1)$; (d) $(1, 0, 1, 0, 1)$. In each case decrypt your ciphertext using the secret information B, $\bar{d}$, h', $\bar{c}$, and A^{-1}; you should get your plaintext back again.

5. Let $q = 2$, $n = 3$, and let $\mathbb{K}$ be represented as the set of polynomials in $\mathbb{F}_2[X]$ modulo the irreducible polynomial $f(X) = X^3 + X + 1$. Use the basis $\{\beta_1, \beta_2, \beta_3\} = \{1, X, X^2\}$; and let $\theta = 2$, $h = 5$, $h' = 3$. Further set

$$A = \begin{pmatrix} 0 & 1 & 0 \\ 1 & 1 & 1 \\ 0 & 0 & 1 \end{pmatrix}, \quad B = \begin{pmatrix} 1 & 0 & 0 \\ 1 & 1 & 0 \\ 0 & 1 & 1 \end{pmatrix}, \quad \bar{c} = (1, 0, 1), \quad \bar{d} = (1, 0, 0) .$$

First express $(u_1 + u_2 X + u_3 X^2)^h$ in terms of the basis with coefficients of the form $\sum u_i u_j$. Then find the public equations for $\bar{y}$ in terms of $\bar{x}$.

§ 2. Patarin's Little Dragon

After breaking the Imai–Matsumoto cryptosystem, Patarin proposed a variant that at first seemed resistant to the type of cryptanalysis in §1.2. After describing this system, we outline an initial attempt at cryptanalysis, using crude linear algebra, that failed. We next give a simple method to break the system if the exponent is not carefully chosen. Finally, we describe the more intricate technique that Coppersmith and Patarin used to break the Little Dragon for any exponent.

2.1 The System

Much of the set-up is the same as in §1.1. As before, $\mathbb{K}$ is an extension of degree n of the finite field $\mathbb{F}_q$, and $\beta_1, \beta_2, \ldots, \beta_n \in \mathbb{K}$ are a basis of $\mathbb{K}$ as an $\mathbb{F}_q$-vector space. Using this basis, Alice, who is preparing to use the Little Dragon cryptosystem in $\mathbb{K}$, regards each element of $\mathbb{K}$ as an n-tuple over $\mathbb{F}_q$. Alice may choose to keep her basis secret, in which case we cannot assume that Catherine the cryptanalyst knows what basis she is using.

Both plaintext message units and ciphertext message units will be n-tuples over $\mathbb{F}_q$. We will use the vector notation $\overline{x} = (x_1, \ldots, x_n) \in \mathbb{F}_q^n$ for plaintext and $\overline{y} = (y_1, \ldots, y_n) \in \mathbb{F}_q^n$ for ciphertext. As before, Alice works with two intermediate vectors $\overline{u} = (u_1, \ldots, u_n) \in \mathbb{F}_q^n$ and $\overline{v} = (v_1, \ldots, v_n) \in \mathbb{F}_q^n$. Given a vector in $\mathbb{F}_q^n$, we use boldface to denote the corresponding element of $\mathbb{K}$ with respect to the basis β_j.

In the Little Dragon cryptosystem, the exponent h has a slightly different form than in the Imai–Matsumoto system. Namely, Alice chooses an exponent h, $0 < h < q^n$, such that $h + 1$ is a sum of two different powers of q, i.e.,

$$h = q^\theta + q^\varphi - 1 , \tag{10}$$

and such that g.c.d.$(h, q^n - 1) = 1$. It is no longer necessary for q to be even. For now we allow Alice to choose the two integers θ and φ arbitrarily in $\{1, \ldots, n-1\}$, subject only to the condition that h be prime to $q^n - 1$. However, in §2.3 we will see that there are some values that give "weak" exponents, i.e., exponents for which the cryptosystem can be readily broken. Alice may choose to keep h secret. But since there are relatively few possibilities for h, she must assume that Catherine the cryptanalyst is prepared to run through all possible h. That is, even if she keeps h secret, the security of her system cannot depend on that.

In addition, Alice chooses two secret linear transformations, i.e., two invertible $n \times n$-matrices $A = \{a_{ij}\}_{1 \leq i, j \leq n}$ and $B = \{b_{ij}\}_{1 \leq i, j \leq n}$ with entries in $\mathbb{F}_q$. *

---

* In the original paper [Patarin 1996b], Alice chooses affine rather than linear transformations. That is somewhat more general, but most likely the added generality does not substantially improve the security of Little Dragon and related systems. In any case, for simplicity we shall assume that the transformations are linear rather than affine.

We now describe how Alice gets her public rule for going from plaintext $\overline{x} \in \mathbb{F}_q^n$ to ciphertext $\overline{y} \in \mathbb{F}_q^n$. First, she sets

$$\overline{u} = A\overline{x} .$$

Next, she would like to have $\mathbf{v} \in \mathbb{K}$ simply equal to the h-th power of $\mathbf{u} \in \mathbb{K}$, and then set

$$\overline{y} = B^{-1}\overline{v} ,$$

where $\overline{v} \in \mathbb{F}_q^n$ is the vector corresponding to $\mathbf{v} \in \mathbb{K}$. But her public encryption rule will go right from $\overline{x}$ to $\overline{y}$ without directly involving exponentiation.

Alice notices that, if $\mathbf{v} = \mathbf{u}^h$, then by (10) she has

$$\mathbf{u}\mathbf{v} = \mathbf{u}^{q^\theta}\mathbf{u}^{q^\varphi} . \tag{11}$$

As in §1.1, she uses the fact that for any $k = 1, 2, \ldots, n$ the operation of raising to the q^k-th power in $\mathbb{K}$ is an $\mathbb{F}_q$-linear transformation. Again let $P^{(k)} = \{p_{ij}^{(k)}\}_{1 \le i,j \le n}$ be the matrix of this linear transformation in the basis $\beta_1, \ldots, \beta_n$ (see equation (2)); and let m_{ijl} be the coefficients when the product $\beta_i\beta_j$ is written as a linear combination of β_l (see equation (3)). Note that (11) can be expanded to give

$$\sum_{1 \le i,j \le n} u_i v_j \beta_i \beta_j = \left(\sum_{i=1}^{n} u_i \beta_i^{q^\theta}\right)\left(\sum_{j=1}^{n} u_j \beta_j^{q^\varphi}\right)$$

$$= \left(\sum_{1 \le i,\mu \le n} p_{i\mu}^{(\theta)} u_i \beta_\mu\right)\left(\sum_{1 \le j,\nu \le n} p_{j\nu}^{(\varphi)} u_j \beta_\nu\right) , \tag{12}$$

by (2). If we use (3) and then compare the coefficients of β_l on the left and right sides of (12), for each l we obtain

$$\sum_{1 \le i,j \le n} m_{ijl} u_i v_j = \sum_{1 \le i,j,\mu,\nu \le n} p_{i\mu}^{(\theta)} p_{j\nu}^{(\varphi)} m_{\mu\nu l} u_i u_j . \tag{13}$$

Of course, Alice knows all of the coefficients m_{ijl} and $p_{ij}^{(k)}$. She now uses her transformations A and B, where

$$\overline{u} = A\overline{x} , \qquad \overline{v} = B\overline{y} ,$$

to replace u_i by $\sum_\rho a_{i\rho} x_\rho$ and replace v_j by $\sum_\sigma b_{j\sigma} y_\sigma$ in (13). When she gathers coefficients of each product $x_i y_j$ and each product $x_i x_j$, she obtains n equations

$$\sum_{1 \le i,j \le n} c_{ijl} x_i y_j + \sum_{1 \le i \le j \le n} d_{ijl} x_i x_j = 0 , \tag{14}$$

$l = 1, 2, \ldots, n$.

Alice makes the equations (14) public. In other words, her public key consists of the $\frac{3}{2}n^3 + \frac{1}{2}n^2$ coefficients c_{ijl}, d_{ijl}. If Bob wants to send her a plaintext message $\overline{x}$, he substitutes the x_i in (14) and solves for the y_j by Gaussian elimination. Here it is crucial that the system (14) is linear in the y_j once the x_i are known. On the other hand, someone who knows only the ciphertext (and the public key) is faced with the daunting task of solving the *nonlinear* system (14) for the unknowns x_i.

When Alice receives the ciphertext $\overline{y}$, she uses her knowledge of A, B, and h to recover $\overline{x}$, without having to solve (14) for the x_i. Let h' be the multiplicative inverse of h modulo $q^n - 1$, so that the map $\mathbf{u} = \mathbf{v}^{h'}$ inverts the map $\mathbf{v} = \mathbf{u}^h$ on $\mathbb{K}$. Alice first computes $\overline{v} = B\overline{y}$, then raises $\mathbf{v} = \sum v_i \beta_i \in \mathbb{K}$ to the h'-th power (i.e., sets $\mathbf{u} = \mathbf{v}^{h'}$), and finally computes $\overline{x} = A^{-1}\overline{u}$.

The following diagram summarizes Alice's decryption:

$$y_1, \ldots, y_n$$
$$\Downarrow$$
$$\overline{v} = B\overline{y}$$
$$\Downarrow$$
$$\mathbf{v} = \sum v_i \beta_i$$
$$\Downarrow$$
$$\mathbf{u} = \mathbf{v}^{h'}$$
$$\Downarrow$$
$$\overline{x} = A^{-1}\overline{u} \ .$$

2.2 A Failed Cryptanalysis

Here is a first attempt to break the Little Dragon cryptosystem.

We won't worry about what basis of $\mathbb{K}$ over $\mathbb{F}_q$ was used by Alice. We'll just use our own convenient basis. Of course, the linear formulas relating $\overline{u}$ to $\overline{x}$ and $\overline{v}$ to $\overline{y}$ are different in our basis, but they're still linear maps. Let A' and B' denote the matrices of these linear transformations. Assume that the exponent h is known (because of its form $h = q^\theta + q^\phi - 1$, there are only a fairly small number of possibilities to be guessed). To break the cryptosystem it suffices to know the matrices A' and B'. We regard the $2n^2$ entries of the matrices A' and B' as unknowns, and we generate a large number of plaintext/ciphertext pairs $(x_1, ..., x_n, y_1, ..., y_n)$. Each such $2n$-tuple can be substituted into the formulas

$$\overline{u} = A'\overline{x} , \qquad \overline{v} = B'\overline{y} ;$$

and the resulting expressions for the u_i and v_j (in terms of the $2n^2$ unknowns) can be substituted into (13) (more precisely, into the equations of the form (13) that we derive using our own basis rather than Alice's basis). Each plaintext/ciphertext pair gives n equations (one for each $l = 1, \ldots, n$) in the $2n^2$ unknowns.

These equations are quadratic, rather than linear. However, if we introduce new variables w_ρ for all of the products of unknowns that appear (i.e., each w_ρ replaces either a product of the form $a_{ij}b_{kl}$ or a product of the form $a_{ij}a_{kl}$), then we obtain linear equations in the $O(n^4)$ new variables. By varying the plaintext/ciphertext pair, we get a vast number of equations in these $O(n^4)$ variables. We then use

Gaussian elimination to find the unknowns w_ρ, and from them it is easy to find the original $2n^2$ unknowns, i.e., the entries in the matrices A' and B'.

Of course, it's crucial to be able to generate enough equations in the $O(n^4)$ unknowns w_ρ, so that the only common solution found by elimination will be the one that's compatible with the fact that each w_ρ is really a product of two of the original $2n^2$ unknowns.

At first glance it seems that, because of the complicated equations (14) used to generate plaintext/ciphertext pairs, we could get enough independent linear equations. However, on closer examination we find that this approach to breaking the system will not work. The reason is that one obtains only $O(n^3)$ independent linear equations in the $O(n^4)$ variables.

To see this, let us look again at the equations that result from (13) after we make the substitutions $\bar{u} = A'\bar{x}$ and $\bar{v} = B'\bar{y}$. After we replace the products $a_{ij}b_{kl}$ and $a_{ij}a_{kl}$ by the corresponding w_ρ, these equations may be regarded as linear equations in the w_ρ whose coefficients are quadratic expressions in $(x_1, \ldots, x_n, y_1, \ldots, y_n)$. More precisely, those coefficients are linear expressions in the $n^2 + n(n+1)/2$ products $x_i y_j$ $(1 \leq i, j \leq n)$ and $x_i x_j$ $(1 \leq i \leq j \leq n)$. Suppose that for each l we construct the following map Φ_l from $\mathbb{F}_q^{n^2+n(n+1)/2}$ to the space of linear equations in the $O(n^4)$ variables w_ρ. To each $\bar{z} \in \mathbb{F}_q^{n^2+n(n+1)/2}$ we associate the linear equation obtained by replacing the n^2 products $x_i y_j$ by the first n^2 components of $\bar{z}$ and the $n(n+1)/2$ products $x_i x_j$ by the remaining components of $\bar{z}$ in the equation in the w_ρ that comes from the l-th equation in (13).

No matter how many plaintext/ciphertext pairs $(x_1, \ldots, x_n, y_1, \ldots, y_n)$ we use, all of the equations in the $O(n^4)$ variables w_ρ that we obtain will be in the image of one of the Φ_l, $l = 1, \ldots, n$. Each image is at most $(n^2+n(n+1)/2)$-dimensional. Thus, the maximum number of independent equations we can possibly hope to generate is $\frac{3}{2}n^3 + \frac{1}{2}n^2$, which is not nearly enough.

2.3 Weak Exponents When $q = 2$

Let $\mathbb{K}$ be an extension of $\mathbb{F}_2$ of degree n. In this section we show how to break Little Dragon if h is a "weak exponent" in the following sense.

In §2.1 we allowed any exponent h of the form $h = 2^\theta + 2^\varphi - 1$ with g.c.d.$(h, 2^n - 1) = 1$. We now suppose that h is such that from the equation

$$\mathbf{v} = \mathbf{u}^h \tag{15}$$

one can obtain an equation of the following form by raising both sides of (15) to some power prime to $2^n - 1$ and multiplying both sides of (15) by powers of $\mathbf{v}$ and $\mathbf{u}$:

$$\mathbf{v}^{2^{i_1}+2^{i_2}+\cdots+2^{i_k}}\mathbf{u}^{2^\alpha} = \mathbf{v}^{2^{j_1}+2^{j_2}+\cdots+2^{j_{k'}}}\mathbf{u}^{2^\beta} , \tag{16}$$

where the number of powers of 2 in the exponents is small (for example, $k, k' \leq 5$).

Example 2.1. Suppose that $n = 6$ and $h = 2^4 + 2^2 - 1 = 19$. Then raising both sides of (15) to the 3rd power and multiplying by $\mathbf{u}^8$ gives $\mathbf{v}^{1+2}\mathbf{u}^{2^3} = \mathbf{u}^{65} = \mathbf{u}^2$, which is of the form (16) with $k = 2$, $k' = 0$, $\alpha = 3$, $\beta = 1$.

In the cryptanalysis, we use the fact that each map $\mathbf{v} \mapsto \mathbf{v}^{2^i \mu}$, $\mathbf{u} \mapsto \mathbf{u}^{2^\alpha}$, $\mathbf{v} \mapsto \mathbf{v}^{2^j \nu}$, and $\mathbf{u} \mapsto \mathbf{u}^{2^\beta}$ is linear. If we follow the same procedure that we used to derive the equations (14) from the relation (11), we see that (16) leads to a set of n equations of the form

$$\sum_{1 \le s_1 \le \cdots \le s_k \le n,\ 1 \le s_0 \le n} e_{s_1,\ldots,s_k,s_0,l}\, y_{s_1} y_{s_2} \cdots y_{s_k}\, x_{s_0}$$

$$= \sum_{1 \le t_1 \le \cdots \le t_{k'} \le n,\ 1 \le t_0 \le n} f_{t_1,\ldots,t_{k'},t_0,l}\, y_{t_1} y_{t_2} \cdots y_{t_{k'}}\, x_{t_0}, \quad (17)$$

$l = 1, 2, \ldots, n$.

Remark. Notice that (16) and (17) are equivalent to one another, i.e., any $(x_1, \ldots, x_n, y_1, \ldots, y_n)$ satisfying (17) gives (by means of the matrices A and B) elements $\mathbf{u}, \mathbf{v} \in \mathbb{K}$ satisfying (16). For $\mathbf{u}, \mathbf{v} \ne 0$ equation (16) is also equivalent to equation (15).

Suppose that Catherine is trying to break Alice's Little Dragon, and knows her exponent h. (As mentioned in §2.1, there are not many possibilities for h, and so Catherine is prepared to run through all possible h.) Suppose that h is "weak", i.e., the relation (15) implies a relation of the form (16). Catherine then knows, first of all, that Alice's plaintext/ciphertext pairs will satisfy a set of at least n equations of the form (17). Second, she knows that, if she finds this set of equations of the form (17), then, by the above remark, for any nonzero n-tuple $(y_1, \ldots, y_n)$ there will be only one nonzero n-tuple $(x_1, \ldots, x_n)$ that satisfies the set of equations (17). Thus, after she finds the equations (17), all that she has to do to decrypt a ciphertext $(y_1, \ldots, y_n)$ is to substitute it into (17) to obtain a *linear* system in the unknowns x_i that has a unique nonzero solution. That solution is the plaintext.

So we have reduced the cryptanalysis to finding all equations of the form (17) that are satisfied by plaintext/ciphertext pairs $(x_1, \ldots, x_n, y_1, \ldots, y_n)$. We regard the coefficients $e_{s_1,\ldots,s_k,s_0,l}$ and $f_{t_1,\ldots,t_{k'},t_0,l}$ as unknowns, and generate a large number of plaintext/ciphertext pairs. For each such $2n$-tuple $(x_1, \ldots, x_n, y_1, \ldots, y_n)$ we obtain a set of n equations (17) that are linear in the unknowns. Without loss of generality we may assume that $k > k'$. Then there are $O(n^{k+2})$ unknowns, and so we expect that after trying $O(n^{k+2})$ different $2n$-tuples $(x_1, \ldots, y_n)$ we will have a complete set of independent equations in the variables $e_{s_1,\ldots,s_k,s_0,l}$ and $f_{t_1,\ldots,t_{k'},t_0,l}$. Using Gaussian elimination, we find the solution space of the equations, i.e., a basis for the space of e- and f-coefficients that give equations satisfied by all plaintext/ciphertext pairs. In other words, we find a maximal set of independent equations of the form (17) that are satisfied by all plaintext/ciphertext pairs. As explained above, this set of equations breaks the cryptosystem, because the equations are linear in the plaintext variables $x_1, \ldots, x_n$.

Remark. The above cryptanalysis in the case of weak exponents works just as well if $\overline{y}$ is related to $\overline{v}$ and $\overline{x}$ is related to $\overline{u}$ by affine rather than linear transformations.

2.4 The Little Dragon is a Paper Tiger:
the Coppersmith–Patarin Cryptanalysis (see [Patarin 1996b])

In this section we show how to break the Little Dragon cryptosystem in the general case.

Let Y be the n-dimensional $\mathbb{F}_q$-vector space of possible ciphertext vectors $\{y_1, \ldots, y_n\}$. Recall that for any vector $\overline{v} = (v_1, \ldots, v_n) \in \mathbb{F}_q^n$ we use boldface to denote the corresponding element of $\mathbb{K}$ with respect to Alice's fixed basis $\beta_1, \ldots, \beta_n$:

$$\mathbf{v} = v_1\beta_1 + \cdots + v_n\beta_n \in \mathbb{K} . \tag{18}$$

Suppose that we somehow managed to stumble upon a bilinear* map that we denote $*$ from $Y \times Y$ to Y:

$$(\overline{y}, \overline{y}') \mapsto \overline{y}'' = \overline{y} * \overline{y}' \tag{19}$$

such that if $\overline{v} = B\overline{y}$, $\overline{v}' = B\overline{y}'$, and $\overline{v}'' = B\overline{y}''$, then $\mathbf{v}'' = \mathbf{v}\mathbf{v}'$. In other words, when the map is translated into $\overline{v}$-vectors using the matrix B it becomes the multiplication map in $\mathbb{K}$. Actually, we shall be satisfied with a map $*$ which has a somewhat weaker property. Namely, we shall be happy if the map $*$ satisfies the following condition: there exists some fixed nonzero $\mu \in \mathbb{K}$ such that for all $\overline{y}$ and $\overline{y}'$, if we apply the matrix B to $\overline{y}$, $\overline{y}'$ and $\overline{y}'' = \overline{y} * \overline{y}'$, then the resulting vectors satisfy

$$\mathbf{v}'' = \mu\, \mathbf{v}\, \mathbf{v}' . \tag{20}$$

Even without knowing B, if we somehow knew that (20) holds, then we could say that an h'-fold iteration of our operation $*$ applied to a ciphertext vector would produce a vector that is related to the plaintext vector $\overline{x}$ by a fixed linear matrix. We now explain this. Let $\overline{y}$ be our ciphertext vector. We define $\overline{y}''$ by setting $\overline{y}' = \overline{y}$ in (19), i.e., $\overline{y}'' = \overline{y} * \overline{y}$; we then define $\overline{y}'''$ to be $\overline{y} * \overline{y}''$; and in general we define

$$\overline{y}^{(j)} = \overline{y} * \overline{y}^{(j-1)} , \qquad j = 2, 3, \ldots, h' .$$

We define

$$\overline{v}^{(j)} = B\overline{y}^{(j)} , \qquad j = 1, 2, \ldots, h' ,$$

and, as always, we let $\mathbf{v}^{(j)}$ denote the element of $\mathbb{K}$ corresponding to $\overline{v}^{(j)}$ as in (18). By applying (20) repeatedly, we find that

$$\mathbf{v}^{(j)} = \mu^{j-1}\mathbf{v}^j , \qquad j = 1, 2, \ldots, h' ;$$

---

* A map from $Y \times Y$ to a vector space is "bilinear" if it is linear in each argument when the other argument is kept fixed.

in particular,

$$\mathbf{v}^{(h')} = \mu^{h'-1}\mathbf{v}^{h'} ,$$

i.e.,

$$\mathbf{u} = \mathbf{v}^{h'} = \mu^{-(h'-1)}\mathbf{v}^{(h')} .$$

Let M be the matrix of multiplication by $\mu^{-(h'-1)} \in \mathbb{K}$ in the basis $\beta_1, \ldots, \beta_n$. Then

$$\overline{x} = A^{-1}\overline{u} = A^{-1}M\overline{v}^{(h')} = A^{-1}MB\overline{y}^{(h')} .$$

In other words, the plaintext vector is equal to the fixed matrix $C = A^{-1}MB$ times the h'-th power (in the sense of the $*$-operation) of $\overline{y}$, as claimed.

So if we knew that (20) holds, we would know that

$$\overline{x} = C\overline{y}^{(h')} \tag{21}$$

for some fixed $n \times n$-matrix C. At that point it would be easy to find the entries of $C = \{c_{ij}\}_{1 \le i,j \le n}$ as follows. We generate a number of plaintext/ciphertext pairs $\overline{x}_l = (x_{0l}, \ldots, x_{nl})$, $\overline{y}_l = (y_{0l}, \ldots, y_{nl})$ for $l = 1, 2, \ldots, L$. It is a simple matter to generate such a pair; in fact, in a public key cryptosystem anyone must be able to encrypt any plaintext of her choosing. In the present situation this is done by arbitrarily choosing the vector $\overline{x}_l$ and then solving the equations (14) (which are linear in the y-variables) for the corresponding ciphertext vector. When we have the ciphertext vector, we find its h'-th power under $*$ (using the repeated squaring method, as in Example 3.5 of Chapter 2), which we then put in the right side of (21). From each plaintext/ciphertext pair $\overline{x}_l, \overline{y}_l$ we get a set of n linear equations (21) in the unknown matrix entries c_{ij}. Once we do this for slightly more than n different $2n$-tuples $(x_{0l}, \ldots, x_{nl}, y_{0l}, \ldots, y_{nl})$, $l = 1, \ldots, L$ with $L > n$, we are almost certain to be able to solve for the n^2 unknowns c_{ij}.

As soon as we know the matrix C, we know how to decrypt using (21), and we have broken Alice's system. Thus, what we need for the cryptanalysis is a bilinear map $* : Y \times Y \longrightarrow Y$ with the desired property. The remainder of this section is devoted to finding such a map $*$.

For each $l = 1, 2, \ldots, n$, let $\delta_l = \delta_l(x_1, \ldots, x_n, y_1, \ldots, y_n)$ denote the first of the two sums in (14), and set $\overline{\delta} = (\delta_1, \ldots, \delta_n)$. The sum δ_l comes from the left side of (13) (which is unknown to Catherine, who doesn't even know Alice's basis $\beta_1, \ldots, \beta_n$) by means of the unknown matrices A and B. That is, the first sum in (14) came from the product $\mathbf{u}\,\mathbf{v}$ on the left in (11).

To create the map $*$, the idea is to exploit the trivial fact that for any $\lambda \in \mathbb{K}$

$$\lambda(\mathbf{u}\mathbf{v}) = \mathbf{u}(\lambda\mathbf{v}) . \tag{22}$$

For every $\lambda \ne 0$ we claim that (22) gives rise to a corresponding pair of $n \times n$-matrices S and T with entries in $\mathbb{F}_q$ such that for all $(x_1, \ldots, x_n, y_1, \ldots, y_n) \in \mathbb{F}_q^{2n}$ the l-th component of $S\overline{\delta}$ is given by

$$(S\overline{\delta})_l = \sum_{1 \le i,j \le n} c_{ijl} x_i (T\overline{y})_j , \tag{23}$$

where $(T\overline{y})_j$ is the j-th component of $T\overline{y}$. Namely, given λ, if we knew the matrix B, and if we knew the matrix Λ of multiplication by λ in Alice's basis $\beta_1, \ldots, \beta_n$, then we would set $T = B^{-1}\Lambda B$ and $S = \Lambda$. This is because $B^{-1}\Lambda B\overline{y}$ is the vector corresponding to $\lambda \mathbf{v}$, and so the right side of (23) would be the β_l-component of the right side of (22). On the left in (23) note that the l-th component of $\Lambda\overline{\delta}$ is the β_l-component of $\lambda(\mathbf{uv}) \in \mathbb{K}$, since the l-th component of $\overline{\delta}$ is the β_l-component of $\mathbf{uv}$.

Of course, we do not know B or Λ. However, what we do know is that such matrices S and T must exist. Moreover, the set of matrices T which have this property (i.e., for which there exists S such that (23) holds) is a vector space of dimension at least n over $\mathbb{F}_q$. Namely, this set contains the set $B^{-1}\Lambda B$ as Λ ranges over the n-dimensional vector space of matrices corresponding to all $\lambda \in \mathbb{K}$. In practice, it seems that the vector space of matrices T is usually n-dimensional, i.e., it usually does not contain anything other than the matrices $B^{-1}\Lambda B$ for $\lambda \in \mathbb{K}$. In what follows, for simplicity we shall assume that the vector space of matrices T for which (23) can be solved for S is of dimension exactly n.

Let the matrices $T_1, \ldots, T_n$ be a basis for this space, so that an arbitrary solution T can be written in the form $T = t_1 T_1 + \cdots + t_n T_n$, where $\overline{t} = (t_1, \ldots, t_n) \in \mathbb{F}_q^n$. A basis of matrices $T_1, \ldots, T_n$ can be found from (23) by Gaussian elimination, where we regard the $2n^2$ entries in the matrices S and T as unknowns and use a large number of plaintext/ciphertext pairs $(x_1, \ldots, x_n, y_1, \ldots, y_n)$ to get as many equations in these unknowns as we need. (See our earlier discussion of how to solve equation (21) for the matrix C.) So from now on we suppose that we have found the matrices $T_1, \ldots, T_n$.

Suppose that we had a function $\overline{t} = f(\lambda)$ from $\mathbb{K}$ to $\mathbb{F}_q^n$ – in other words, an n-tuple of functions $t_i = f_i(\lambda)$ from $\mathbb{K}$ to $\mathbb{F}_q$ – that gives us the T corresponding to λ, i.e., that satisfies $\sum f_i(\lambda)T_i = B^{-1}\Lambda B$ for all $\lambda \in \mathbb{K}$, where Λ denotes the matrix of multiplication by λ in the basis $\beta_1, \ldots, \beta_n$. Such a function f would give a linear map (in fact, a vector space isomorphism) between $\mathbb{K}$ and the space of solutions T. Now let g_i be the map from vectors $\overline{y}$ to $\mathbb{F}_q$ that takes $\overline{y}$ to $\overline{v} = B\overline{y}$ and then applies f_i to $\mathbf{v} = \sum v_j\beta_j$:

$$g_i : \overline{y} \mapsto \overline{v} = B\overline{y} \mapsto \mathbf{v} = \sum_{i=j}^{n} v_j\beta_j \mapsto t_i = f_i(\mathbf{v}) .$$

If we knew g, we could define our operation $*$ as follows:

$$\overline{y}'' = \overline{y} * \overline{y}' = \sum_{i=1}^{n} g_i(\overline{y})T_i\overline{y}' . \tag{24}$$

Then we would have $\mathbf{v}'' = \mathbf{v}\mathbf{v}'$, where, as always, $\mathbf{v}$ denotes the element $\sum v_i\beta_i$ of $\mathbb{K}$ corresponding to $\overline{v} = B\overline{y}$, and similarly for $\mathbf{v}'$ and $\mathbf{v}''$.

However, as remarked before, we do not really need $\mathbf{v}'' = \mathbf{v}\mathbf{v}'$; it suffices to have (20). Thus, we will be satisfied with a linear map f that satisfies a more general property. Namely, for an arbitrary fixed nonzero $\mu \in \mathbb{K}$ and for every

$\lambda \in \mathbb{K}$ let $\widetilde{\Lambda}$ denote the matrix of multiplication by $\mu\lambda$ in the basis $\beta_1, \ldots, \beta_n$. Suppose that $\bar{t} = f(\lambda)$ is a linear map from $\mathbb{K}$ to $\mathbb{F}_q^n$ such that for some fixed μ one has $\sum f_i(\lambda)T_i = B^{-1}\widetilde{\Lambda}B$ for all $\lambda \in \mathbb{K}$; and let $g(\overline{y})$ be the map obtained by composing this f with B as before. If we knew such a g, then all we would have to do is define $*$ by (24) with this g. We would then have $\mathbf{v}'' = \mu\,\mathbf{v}\,\mathbf{v}'$, which is the relation (20) that we need.

How do we find such a linear map g? We use the crucial but obvious fact that any operation $*$ satisfying (20) is commutative. Let $G = \{g_{ij}\}$ be the matrix of g: $g(\overline{y}) = G\overline{y}$. Let G_i denote the i-th row of G. If we can find G, then we define $*$ by setting

$$\overline{y} * \overline{y}' = \sum_{i=1}^{n} G_i\overline{y}\,T_i\overline{y}' \,. \tag{25}$$

We regard the entries g_{ij} in G as unknowns, and we use the fact that the operation in (25) is commutative, i.e.,

$$\sum_{i=1}^{n} G_i\overline{y}T_i\overline{y}' = \sum_{i=1}^{n} G_i\overline{y}'T_i\overline{y} \,. \tag{26}$$

Let $(T_i)_{\sigma\tau}$ denote the $\sigma\tau$-entry of the matrix T_i. For $1 \leq j_1, j_2, k_0 \leq n$ we choose $\overline{y}$ to be the j_1-th standard basis vector, choose $\overline{y}'$ to be the j_2-th standard basis vector, and compare the k_0-th component of the vector equation (26). We obtain

$$\sum_{i=1}^{n} G_{ij_1}(T_i)_{k_0j_2} = \sum_{i=1}^{n} G_{ij_2}(T_i)_{k_0j_1} \,.$$

This gives us n^3 equations in the unknowns G_{ij}. We know that there is at least an n-dimensional solution space – since every fixed $\mu \in \mathbb{K}$ gives a matrix G – and in practice it is not likely for there to be other solutions G that do not come about in this way. All we need to find is any nonzero solution G in this n-dimensional space of solutions. Once we have such a G, we define the $*$ operation by (25), and then, as explained before, we can break Alice's cryptosystem. This concludes our description of the Coppersmith–Patarin cryptanalysis of Little Dragon.

Exercises for § 2

1. Show that an equation of the form (16) with q in place of 2 cannot be obtained without the assumption that $q = 2$.
2. Show that $h = 2^{n-2} + 2^{n-3} - 1$ is a weak exponent with the same k and k' as in Example 2.1.

§3. Systems That Might Be More Secure

Patarin investigated several generalizations and extensions of the Imai–Matsumoto system that in some cases appear to resist attacks such as the ones in §1.2 and §2.4. In this section we discuss some of these systems.

3.1 Big Dragon

$\mathbb{K}$ is an extension of degree n of the finite field $\mathbb{F}_q$ of characteristic 2, and $\beta_1, \beta_2, \ldots, \beta_n \in \mathbb{K}$ form a basis of $\mathbb{K}$ as an $\mathbb{F}_q$-vector space. Alice may keep her basis secret, if she chooses. As usual, by means of the basis she thinks of each element of $\mathbb{K}$ as an n-tuple over $\mathbb{F}_q$. We use boldface for an element of $\mathbb{K}$ and overlining for the corresponding n-tuple.

Again $\overline{x} = (x_1, \ldots, x_n) \in \mathbb{F}_q^n$ denotes plaintext, $\overline{y} = (y_1, \ldots, y_n) \in \mathbb{F}_q^n$ denotes ciphertext, and $\overline{u} = (u_1, \ldots, u_n) \in \mathbb{F}_q^n$ and $\overline{v} = (v_1, \ldots, v_n) \in \mathbb{F}_q^n$ are two intermediate vectors. These intermediate vectors are related to $\overline{x}$ and $\overline{y}$ as in (6), where the matrices A and B and the fixed vectors $\overline{c}$ and $\overline{d}$ are secret.

Alice now chooses an integer h of the form

$$h = q^{\theta_1} + q^{\theta_2} - q^{\varphi_1} - q^{\varphi_2} \tag{27}$$

such that g.c.d.$(h, q^n - 1) = 1$. She chooses a secret $\mathbb{F}_q$-linear map $\psi : \mathbb{K} \to \mathbb{K}$. (One might want to allow ψ to be affine rather than linear.) The relation between $\mathbf{u}$ and $\mathbf{v}$ is that

$$\mathbf{u}^h = \frac{\psi(\mathbf{v})}{\mathbf{v}} \qquad \text{for} \qquad \mathbf{u}, \mathbf{v} \in \mathbb{K}, \ \mathbf{v} \neq 0 \ . \tag{28}$$

Equivalently, for $\mathbf{u}, \mathbf{v} \in \mathbb{K}$ we want to have

$$\mathbf{u}^{q^{\theta_1}+q^{\theta_2}} \mathbf{v} = \mathbf{u}^{q^{\varphi_1}+q^{\varphi_2}} \psi(\mathbf{v}) \ . \tag{29}$$

Since we want the correspondence between $\mathbf{u}$ and $\mathbf{v}$ to be a bijection, ψ must be chosen so that the map $\mathbf{v} \mapsto \psi(\mathbf{v})/\mathbf{v}$ is one-to-one on the set $\mathbb{K}^*$ of nonzero elements of $\mathbb{K}$.

Example 3.1. If $q = 2$, α is an integer such that g.c.d.$(\alpha, n) = 1$, and $\mu \in \mathbb{K}^*$, then the map

$$\psi(\mathbf{v}) = \mu \, \mathbf{v}^{q^\alpha}$$

has the required property. (See Exercise 1 of §1.) Here Alice keeps α and μ secret.

Example 3.2. If $\mu, \nu \in \mathbb{K}^*$ are any secret elements, then the affine map

$$\psi(\mathbf{v}) = \mu \, \mathbf{v} + \nu$$

has the required property.

Recall how in §1 Alice came up with n polynomial equations of degree 2 in the x_i and degree 1 in the y_i, starting from the relation (1). She used the coefficients expressing $\beta_i^{q^k}$ in terms of a basis, expressing $\beta_i\beta_j$ in terms of a basis, and expressing $\overline{x}$ and $\overline{y}$ in terms of $\overline{u}$ and $\overline{v}$ (see (2), (3) and (6), respectively). If we proceed in the same way starting with the equation (29), we arrive at n polynomial equations (one for each basis element) of total degree 3 in the variables $\{x_1, \ldots, x_n, y_1, \ldots, y_n\}$ but only of degree 1 in the y_i. That is, a typical term would have the form $c_{ijk}x_ix_jy_k$, and there might also be x_iy_k-terms, x_ix_j-terms, y_k-terms, and so on. As in §1 and §2, Alice makes these n equations public. If Bob wants to send her a plaintext message unit $\overline{x}$, he substitutes those plaintext coordinates in the equations and solves the resulting *linear* system for the ciphertext $\overline{y}$. Given a ciphertext $\overline{y}$, the intruder Catherine is confronted with a set of *nonlinear* equations for the x_i. Alice, on the other hand, can use equation (28) to decipher. That is, she uses (6) to transform $\overline{y}$ to $\overline{v}$, then computes

$$\mathbf{u} = (\psi(\mathbf{v})/\mathbf{v})^{h'} ,$$

where h' is the inverse of h modulo $q^n - 1$. Finally, she again uses (6) to transform $\overline{u}$ to $\overline{x}$.

Unfortunately, as explained in Patarin's expanded version of [1996b], the Big Dragon is often vulnerable to the same type of attack as Little Dragon (see §2.4), at least when the function $\psi(\mathbf{v})$ is publicly known. If $\psi(\mathbf{v})$ is kept secret, however, it is not clear how to attack the system. Even in that case one must be cautious, because the system is very new. Until a large number of people have spent a lot of time trying to break the Big Dragon with secret ψ, we cannot have confidence in its security.

3.2 Double-Round Quadratic Enciphering (see [Goubin and Patarin 1998a])

Let $\mathbb{K}$ be an extension of degree n of $\mathbb{F}_q$, where n is odd and $q \equiv 3 \pmod 4$. As before, $\beta_1, \beta_2, \ldots, \beta_n \in \mathbb{K}$ form an $\mathbb{F}_q$-basis of $\mathbb{K}$. We suppose that this basis is publicly known; however, Alice will later introduce secret matrices in order to disguise the identification of $\mathbb{K}$ with $\mathbb{F}_q^n$. As before, we use boldface for elements of $\mathbb{K}$ and overlining for the corresponding n-tuples. Thus, if $\overline{x} = (x_1, \ldots, x_n) \in \mathbb{F}_q^n$, then $\mathbf{x} = x_1\beta_1 + \cdots + x_n\beta_n \in \mathbb{K}$.

Alice chooses three secret invertible $n \times n$-matrices A, B, and C with entries in $\mathbb{F}_q$. To transform plaintext $\overline{x}$ into ciphertext $\overline{y}$, she uses four intermediate vectors $\overline{u}$, $\overline{v}$, $\overline{w}$, $\overline{z}$. She successively sets

$$\overline{u} = A\overline{x} , \qquad \mathbf{v} = \mathbf{u}^2 , \qquad \overline{w} = B\overline{v} , \qquad \mathbf{z} = \mathbf{w}^2 , \qquad \overline{y} = C\overline{z} .$$

Using the coefficients m_{ijl} in (3), along with the entries in A and B, Alice can express each w_l as a homogeneous quadratic polynomial in the n variables $x_1, \ldots, x_n$. In other words, she can obtain relations of the form $w_l = \sum_{1 \leq i \leq j \leq n} \alpha_{ijl}x_ix_j$, $l = 1, \ldots, n$, where $\alpha_{ijl} \in \mathbb{F}_q$. Similarly, she can express $\overline{y}$ in terms of $\overline{w}$ using n homogeneous quadratic polynomials in the n variables $w_1, \ldots, w_n$. Composing these two maps, she finally obtains n polynomials

$$y_l = p_l(x_1, \ldots, x_n)$$

that are homogeneous of degree 4. Alice's public key consists of the polynomials p_l. Her private key is the triple of matrices A, B, C.

There is one minor problem with this cryptosystem: the squaring map from $\mathbb{K}^*$ to $\mathbb{K}^*$ is not bijective, but rather is 2-to-1. That is, both $\overline{x}$ and $-\overline{x}$ give the same ciphertext. However, it is not hard to straighten this out by slightly modifying the message space in which we take the plaintext and ciphertext. We now show how to make this modification.

Let the message space $\mathbb{M}$ be a convenient set of representatives modulo ± 1 of the nonzero vectors in $\mathbb{F}_q^n$. For example, if $\mathbb{F}_q = \mathbb{F}_p$ is a prime field, choose $\mathbb{M}$ to be the set of elements whose first nonzero component x_i is between 1 and $(p-1)/2$. If $\mathbb{F}_q$ is an extension of degree > 1 of a prime field $\mathbb{F}_p$, then write elements of $\mathbb{F}_q$ in terms of a fixed $\mathbb{F}_p$-basis, and define $\mathbb{M}$ to be the set of nonzero vectors $\overline{x} \in \mathbb{F}_q^n$ whose first nonzero component x_i has the property that its first nonzero component in the $\mathbb{F}_p$-basis falls between 1 and $(p-1)/2$.

For any nonzero $\overline{x} \in \mathbb{F}_q^n$, exactly one of the elements $\{\overline{x}, -\overline{x}\}$ is in $\mathbb{M}$. We shall write $\pm\overline{x}$ to denote whichever of these two elements belongs to $\mathbb{M}$. The linear maps given by the matrices A, B, and C may be regarded as maps from $\mathbb{M}$ to $\mathbb{M}$. We shall write, for example, $\pm\overline{u} = A(\pm\overline{x}) = \pm A\overline{x}$.

If an element $\mathbf{u} \in \mathbb{K}^*$ has the property that the corresponding vector $\overline{u}$ belongs to $\mathbb{M}$, then we shall also write $\mathbf{u} \in \mathbb{M}$; in this way $\mathbb{M}$ may be regarded as a subset of $\mathbb{K}^*$. For any $\mathbf{u} \in \mathbb{K}^*$ we write $\pm\mathbf{u}$ to denote whichever of $\mathbf{u}$ or $-\mathbf{u}$ belongs to $\mathbb{M}$.

The squaring maps $\mathbf{v} = \mathbf{u}^2$ and $\mathbf{z} = \mathbf{w}^2$, when considered as maps from $\mathbb{M}$ to $\mathbb{M}$, are bijections. This is because -1 is a non-square in $\mathbb{K}$ (here we are using the assumption that $q \equiv 3 \pmod 4$ and n is odd), and so for any $\mathbf{x} \in \mathbb{K}^*$ exactly one of the two elements $\mathbf{x}$ and $-\mathbf{x}$ is a square.

To summarize, Alice gives her encryption map from $\mathbb{M}$ to $\mathbb{M}$ in the form of n degree-4 polynomials in n variables $\overline{x} = (x_1, \ldots, x_n)$:

$$\pm\overline{y} = \pm\overline{p}(\overline{x}) \, , \qquad \overline{p}(\overline{x}) = (p_1(\overline{x}), \ldots, p_n(\overline{x})) \, ,$$

which she computed by successively applying the maps

$$\pm\overline{x} \mapsto \pm\overline{u} = \pm A\overline{x} \mapsto \mathbf{v} = (\pm\mathbf{u})^2 \mapsto$$
$$\mapsto \overline{w} = B\overline{v} \mapsto \mathbf{z} = (\pm\mathbf{w})^2 \mapsto \pm\overline{y} = \pm C\overline{z} \, .$$

To decrypt a message $\pm\overline{y}$, Alice uses the fact that raising to the $\left(\frac{q^n+1}{4}\right)$-th power inverts the squaring map. Namely, if we have $\mathbf{v} = (\pm\mathbf{u})^2$, then

$$\mathbf{v}^{(q^n+1)/4} = \mathbf{u}^{(q^n+1)/2} = \mathbf{u}^{(q^n-1)/2} \cdot \mathbf{u} = \pm\mathbf{u} \, .$$

(Again we're using the assumption that $q \equiv 3 \pmod 4$ and n is odd. We shall return to the subject of computing square roots in $\mathbb{F}_q$ in more generality in §1.8 of Chapter 6.) Thus, Alice goes from $\pm\overline{y}$ to $\pm\overline{x}$ as follows:

$$\pm \overline{y} \mapsto \pm \overline{z} = C^{-1}(\pm \overline{y}) \mapsto \pm \mathbf{w} = (\pm \mathbf{z})^{(q^n+1)/4} \mapsto$$

$$\mapsto \pm \overline{v} = B^{-1}(\pm \overline{w}) \mapsto \pm \mathbf{u} = (\pm \mathbf{v})^{(q^n+1)/4} \mapsto \pm \overline{x} = A^{-1}(\pm \overline{u}) .$$

Catherine, who is trying to break the cipher without knowing A, B, or C, is faced with the problem of solving a system of n degree-4 equations in the n unknowns $x_1, \ldots, x_n$.

Why are two rounds necessary? Couldn't we simply set $\pm \overline{y} = \pm B \overline{v}$, in which case each $\pm y_l$ would be given by a quadratic polynomial in $x_1, \ldots, x_n$? It turns out that such a one-round quadratic encryption is insecure. Goubin and Patarin [1998a] show this as follows. Let $\overline{y} = f(\overline{x})$ be given by the composition of the three maps

$$\overline{u} = A \overline{x} , \qquad \mathbf{v} = \mathbf{u}^2 , \qquad \overline{y} = B \overline{v} , \tag{30}$$

where A and B are secret. Consider the following function in $2n$ variables $x_1, \ldots, x_n, x_1', \ldots, x_n'$:

$$\varphi(\overline{x}, \overline{x}') = \frac{1}{4} \Big(f(\overline{x} + \overline{x}') - f(\overline{x} - \overline{x}') \Big) .$$

From (30) it follows that $\varphi(\overline{x}, \overline{x}')$ is bilinear; that is, it has the form $\varphi(\overline{x}, \overline{x}') = \sum \alpha_{ij} x_i x_j$. Namely, if we let $\overline{u} = A \overline{x}$ and $\overline{u}' = A \overline{x}'$, we see that $B^{-1} \varphi(\overline{x}, \overline{x}')$ is the vector corresponding to the following element of $\mathbb{K}$:

$$\frac{1}{4} \Big((\mathbf{u} + \mathbf{u}')^2 - (\mathbf{u} - \mathbf{u}')^2 \Big) = \mathbf{u} \mathbf{u}' .$$

Since $\mathbf{u} \mathbf{u}'$ is a bilinear function of $\overline{x}, \overline{x}'$, so is $\varphi(\overline{x}, \overline{x}')$.

Using $\varphi(\overline{x}, \overline{x}')$, Catherine can break the system. The way she does this is very similar to the cryptanalysis in §2.4. See Exercise 4 below.

At present the double-round quadratic encryption has not been broken. One possible approach to cryptanalyzing it is to try to separate the degree-4 map $\pm \overline{y} = \pm \overline{p}(x_1, \ldots, x_n)$ into its two quadratic components. That is, one would want to solve the following problem: Let $p_l(x_1, \ldots, x_n)$, $l = 1, \ldots, n$, be homogeneous degree-4 polynomials in n variables. Suppose that there exist quadratic polynomials $\widetilde{p}_l(w_1, \ldots, w_n)$, $l = 1, \ldots, n$, and $w_l(x_1, \ldots, x_n)$, $l = 1, \ldots, n$, such that $p_l(x_1, \ldots, x_n) = \widetilde{p}_l(w_1(x_1, \ldots, x_n), \ldots, w_n(x_1, \ldots, x_n))$ for $l = 1, \ldots, n$. Find an algorithm that computes the $\widetilde{p}_l$ and w_l if one is given the p_l.

This is a special case of the general problem of functional decomposition of polynomials, where one tries to express multivariate polynomial functions as a composition of polynomials of lower degree. Such problems are usually quite intractable; the running times for algorithms to solve them tend to grow exponentially with the number n of variables. See [Dickerson 1989] and [von zur Gathen 1990a and 1990b].

In [1998a] Goubin and Patarin compare the running times of the double-round quadratic cryptosystem with those of other cryptosystems, such as RSA. They conclude that Alice's decryption goes much faster, and so lends itself to efficient implementation on smart cards. However, more work has to be done before one can have confidence in the security of the system.

3.3 Signatures

From §§2, 3, and 6 of Chapter 1 recall that a hash function mapping a long message M to a much shorter sequence of symbols H must have the following property. It is computationally infeasible for Catherine to tamper with M in such a way as to create a new message with the same hash value H. Also recall from Chapter 1 that a digital signature of a message from Bob to Alice means the following. Bob performs an operation $H \mapsto \overline{x}$ that Alice is sure could only have been performed by Bob. Bob appends this signature $\overline{x}$ to the plaintext message M before encrypting the whole thing and sending it to Alice.

Suppose that Bob wants to set up a digital signature system that gives signatures that are as short as possible – say, about 64 bits. He is willing to do a fair amount of computation to calculate each signature, but he is severely limited in the size of the signature. Here is Patarin's proposed system for short signatures.

We suppose that Bob, like Alice before, is working over a field $\mathbb{K}$ of degree n over $\mathbb{F}_q$. We further suppose that q is small (possibly $q = 2$). We let the hash function take values in an m-dimensional $\mathbb{F}_q$-vector space V. In practice, m will usually be greater than n (for example, $m \approx 2n$).

The procedure described below will not work for all possible hash values $H \in V$, but only for about 63% of them (see Exercise 5 at the end of this section; here we are supposing that f_H in (31) behaves like a random polynomial as H varies). Bob needs to have several different values for H, so that he can be virtually certain that the procedure will work for at least one of them. For instance, if he has a list of 26 hash values for his message, then the probability will be only about one in 200 billion that the procedure will not work for any of the values. One way to get 26 different hash values is to apply the hash function to the message with a single letter A through Z appended. Of course, Alice knows that she should accept a signature for any one of the 26 hash values that result from such a modified message.

Rather than a monomial $\mathbf{u}^h$, Bob constructs a polynomial in $\mathbf{u}$ whose degree d is less than some reasonable bound (Patarin suggests $d \leq 8000$). The powers of $\mathbf{u}$ are all either a power of q or a sum of two powers of q, and the coefficients are affine $\mathbb{F}_q$-vector space functions $\psi_i : V \to \mathbb{K}$. The powers of $\mathbf{u}$ and the coefficient functions ψ_i are kept secret. In other words, Bob selects a secret degree-d polynomial

$$f_H(\mathbf{u}) = \psi_0(H) + \sum_{i=1}^{k} \psi_i(H)\mathbf{u}^{q^{\alpha_i}} + \sum_{i=k+1}^{l} \psi_i(H)\mathbf{u}^{q^{\beta_i}+q^{\gamma_i}} . \tag{31}$$

Let $\mathbf{u} = A\overline{x}+\overline{c}$, as in (6); and let $y_1, \ldots, y_m$ denote the components of a vector $H \in V$. Bob uses the relations (2), (3) and (6) to transform f_H to n polynomials $F_i(\overline{x}, \overline{y}) \in \mathbb{F}_q[x_1, \ldots, x_n, y_1, \ldots, y_m]$ (one for each element of Bob's fixed $\mathbb{F}_q$-basis of $\mathbb{K}$), each of which has total degree 2 in the variables $x_1, \ldots, x_m$ and total degree 1 in the variables $y_1, \ldots, y_m$. Bob makes the F_i public.

When Bob wants to sign a hash value $H = (y_1, \ldots, y_m)$, he substitutes this value into the coefficients $\psi_i(H)$ and considers the resulting degree-d polynomial $f_H \in \mathbb{K}[\mathbf{u}]$. He finds a root of the equation $f_H = 0$, provided that this polynomial has a root in $\mathbb{K}$. Below we will describe an efficient algorithm for finding a root, if there is one. If there is no root, then he starts trying the other 25 values of H, until he finds one for which the equation $f_H = 0$ has a root in $\mathbb{K}$. Let $\mathbf{u}$ be such a root. Using (6), he transforms $\mathbf{u}$ to a vector $\overline{x}$, which he sends to Alice. This $\overline{x} \in \mathbb{F}_q^n$ is his signature.

Alice's task now is pretty simple. She computes the 26 hash values of the message with an appended letter. For each such value $H = (y_1, \ldots, y_m)$ she checks whether or not $F_i(x_1, \ldots, x_n, y_1, \ldots, y_m) = 0$, $i = 1, \ldots, n$. If any of the 26 values of H combines with the signature $\overline{x}$ to give a solution of this system of equations, then Alice knows that Bob really did send the message, and the message has not been tampered with. Otherwise, she knows that Catherine has been causing trouble.

Notice that if Catherine tries to impersonate Bob and send her own message with hash value $H = (y_1, \ldots, y_m)$, then to find a signature vector $\overline{x}$ she has to solve a system of n equations of degree 2.

Because the ψ_i are affine functions, Catherine would be able to solve the equations for $\overline{y}$ given any $\overline{x}$. Because $m > n$, she could actually find a large number of $\overline{y}$ corresponding to each $\overline{x}$. But such $\overline{y}$ would not do her much good, since she'd be unable to reverse the hash function to find a message with hash value $\overline{y}$. If for some reason we wanted to prevent Catherine even from finding $\overline{y}$, we could replace the affine ψ_i by higher-degree polynomials. However, from a practical standpoint there does not seem to be any reason to make the coefficients more complicated in such a way.

Finally, we explain how Bob can find a root $\mathbf{u} \in \mathbb{K}$ of the polynomial $f_H(X) \in \mathbb{K}[X]$ in (31) if it has such a root, that is, if the polynomial $f(X) = $ g.c.d.$(f_H(X), X^{q^n} - X)$ has degree greater than zero. Let η be a randomly chosen nonzero element of $\mathbb{K}$. Consider the polynomial $g(X) = g_\eta(X) = \sum_{i=0}^{n-1} (\eta X)^{q^i}$. The polynomial $g_1(X) = \sum X^{q^i}$ is the trace map. By Exercise 13 in §2 of Chapter 3, for fixed η the polynomial $g(X)$ takes any given value $c \in \mathbb{F}_q$ for exactly q^{n-1} different $X \in \mathbb{K}$. For any $c \in \mathbb{F}_q$ the polynomial

$$\widetilde{f}(X) = \text{g.c.d.}(f(X), g(X) - c) \tag{32}$$

is equal to the product of $(X - \mathbf{u})$ over all roots $\mathbf{u} \in \mathbb{K}$ of $f_H(X)$ such that $\eta\mathbf{u}$ has trace c.

Bob varies η randomly, and makes the g.c.d. computation in (32) and the analogous computations with $f(X)$ replaced by the factors of $f(X)$ that are split off by the earlier g.c.d. computations. Although occasionally he might get a trivial g.c.d. – either 1 or $f(X)$ – it can be shown that he is almost certain to be able to progressively split off factors of $f(X)$ in $\mathbb{K}[X]$, until he finally obtains a factor of the form $X - \mathbf{u}$. For more details of this algorithm, along with other methods of finding roots of polynomials over finite fields, see [Lidl and Niederreiter 1986].

Exercises for §3

1. Let $\mathbb{K} = \mathbb{F}_{2^n}$. Show that the squaring map is bijective. But explain why the cryptosystem in §3.2 is completely insecure when $q = 2$, and in fact when q is any power of 2.

2. Is the squaring map bijective on $GL_2(\mathbb{F}_{2^n})$? Explain. (Here $GL_2(\mathbb{K})$ denotes the set of invertible 2×2-matrices with entries in $\mathbb{K}$.)

3. If q were $\equiv 1 \pmod 4$ or if n were even in §3.2, show that the encryption map $\overline{x} \mapsto \overline{u} \mapsto \overline{v} \mapsto \overline{w} \mapsto \overline{z} \mapsto \overline{y}$ would be 4-to-1 rather than 2-to-1.

4. Using the bilinear map $\varphi(\overline{x}, \overline{x}')$ in the text and proceeding as in §2.4, show how to break the one-round quadratic enciphering (30).

5. Show that for a large finite field $\mathbb{K} = \mathbb{F}_{q^n}$ and large d, the proportion of monic degree-d polynomials $f(X)$ having a root in $\mathbb{K}$ is very close to $1 - \frac{1}{e} \approx 63.2\%$. Do this in two ways:

(a) Make the heuristic assumption that such polynomials may be regarded as random functions from $\mathbb{K}$ to $\mathbb{K}$. For each $x \in \mathbb{K}$ there is a $1 - (1/q)$ probability that the value of such a function is nonzero. Then compute the probability that $f(x) = 0$ for some $x \in \mathbb{K}$.

(b) Without making the heuristic assumption in part (a), work directly with polynomials. Use the fact that the number of monic degree d polynomials that are divisible by $(x - x_1)(x - x_2) \cdots (x - x_r)$ is q^{d-r}.

Chapter 5. Combinatorial–Algebraic Cryptosystems

§ 1. History

About twenty years ago, a combinatorial cryptosystem called the Merkle–Hellman Knapsack met with a great deal of enthusiastic acclaim [Hellman and Merkle 1978]. For message transmission it was much more efficient than its main competitor at the time, which was RSA. Moreover, it was thought to be almost *provably* secure. Whereas the security of RSA is based on the difficulty of factoring large integers, that of Merkle–Hellman is based on the conjecturally more difficult Subset Sum problem, which is known to be NP-complete (see Definition 4.6 of Chapter 2).

But within a few years Adi Shamir [1984] completely broke Merkle–Hellman, by showing that the *subproblem* (special case) of Subset Sum that its security relies upon can be solved in polynomial time. (This did not, of course, disprove the P≠NP conjecture, since Shamir's algorithm was only for a subproblem.) Although generalizations and modifications of Merkle–Hellman were introduced in an attempt to salvage the situation, in the 1980's most of them were also broken by Shamir, Brickell, Lagarias, Odlyzko and others (see [Brickell 1985] and [Odlyzko 1990]). This painful experience traumatized many cryptographers, and partly for this reason combinatorially based cryptosystems fell into disfavor.

There was a second reason for the pessimism about combinatorial cryptography: Brassard's theorem [Brassard 1979]. This theorem, in the words of Odlyzko [1990], "says essentially that if breaking a cryptosystem is NP-hard [see Definition 4.7 of Chapter 2], then NP=co-NP [see Definition 4.2 of Chapter 2], which would be a very surprising complexity theory result". The common interpretation of Brassard's theorem was that cryptography must be based not on NP-complete problems – which include most of the interesting problems of combinatorics – but rather on problems that are thought to be of intermediate difficulty (that is, strictly between P and NP-complete), such as factoring large integers or finding discrete logarithms in a finite field. This feeling was summarized in an important article by Selman [1988], who said: "There can be no hope to transform arbitrary problems in NP–P into public key cryptosystems."

However, we now have some evidence that this verdict condemning combinatorial cryptography might have been premature. In the first place, there is less in Brassard's theorem than meets the eye. Namely, the central hypothesis of the theorem seems not to hold for the combinatorial constructions that have recently been proposed for public key cryptosystems. In the second place, in [Fellows and Koblitz 1994b] we show how to generate an entire class of hybrid combinatorial-and-algebraic cryptosystems.

§ 2. Irrelevance of Brassard's Theorem

In a public key cryptosystem there are two types of one-way functions:

1) the encryption function (whose inversion is the *cracking problem* – see §5.1 of Chapter 2); and
2) the underlying function used to construct the trapdoor (see §5.2 of Chapter 2).

In combinatorial cryptography it is the second of these that uses a basic problem in combinatorics. It is also the second type of function that was considered in [Brassard 1979]. So in what follows we shall use the term "one-way function" in the sense 2). (When we first discussed this term in §6 of Chapter 1, we understood it in the sense 1).)

In [Brassard 1979] the two examples given were for the RSA (factoring) and Diffie–Hellman (discrete log) cryptosystems. In RSA the one-way function φ is:

$$\varphi: \ \mathcal{P} \times \mathcal{P} \xrightarrow{\text{multiply}} \mathbb{N}$$

($\mathcal{P}$=set of primes, $\mathbb{N}$=set of natural numbers). In Diffie–Hellman the one-way function φ is

$$\varphi: \ \mathbb{Z}/(q-1)\mathbb{Z} \xrightarrow{\text{exponentiate to base } g} \mathbb{F}_q \ .$$

(g is a fixed element of the finite field $\mathbb{F}_q$).

Brassard's theorem assumes the following condition:

$$\text{image}(\varphi) \in \text{co-NP} \ . \tag{1}$$

That is, there must exist a polynomial time certificate for something not being in the image of φ (see Definition 4.2 of Chapter 2). This hypothesis tends to hold for number-theoretic one-way functions. For example, it is not hard to show that there exists a polynomial time certificate that $n \in \mathbb{N}$ is a prime or a product of three or more primes, or that $y \in \mathbb{F}_q^*$ is not in the subgroup generated by g. But most likely the assumption (1) does *not* hold for most combinatorial one-way functions.

Example 2.1. Reversible cellular automata ([Kari 1992]; see also [Guan 1987] and [Wolfram 1986]). A d-dimensional cellular automaton is defined as follows. Let S be a finite set whose elements are called "states", and let $\bar{a}_1,\ldots,\bar{a}_n \in \mathbb{Z}^d$ be a fixed set of vectors. A "neighborhood" of a vector $\bar{x} \in \mathbb{Z}^d$ is the set $\{\bar{x}+\bar{a}_1,\ldots,\bar{x}+\bar{a}_n\}$. A "configuration" C is a map from $\mathbb{Z}^d$ to S, i.e., an assignment of states to vectors (or "cells"). A cellular automaton is defined by a "local rule" $f: \ S^n \to S$. Such a local rule determines a map $\mathcal{A}$ from one configuration to another as follows: $\mathcal{A}(C)(\overline{x}) = f(C(\overline{x} + \overline{a}_1), \ldots, C(\overline{x} + \overline{a}_n))$. In other words, the state of the configuration $\mathcal{A}(C)$ at the cell $\overline{x}$ depends only on the states of C at the neighboring cells $\overline{x} + \overline{a}_i$ in a manner described by f. A cellular automaton is said to be "reversible" if $\mathcal{A}$ is injective, that is, if every configuration C' can be uniquely retraced back one step to a configuration C such that $\mathcal{A}(C) = C'$.

In Kari's cryptosystem, $\{\mathcal{A}_i\}$ is a set of easy-to-invert reversible cellular automata. The one-way function φ is composition of cellular automata:

$$\varphi : (\mathcal{A}_{i_1}, \ldots, \mathcal{A}_{i_\ell}) \mapsto \mathcal{A} = \mathcal{A}_{i_1} \circ \cdots \circ \mathcal{A}_{i_\ell} .$$

It is hard to imagine what polynomial time certificate could exist that would show that a given reversible cellular automaton cannot be written as a composition of the $\mathcal{A}_i$.

Example 2.2. Rewrite systems [Do Long Van, Jeyanthi, Siromoney, and Subramanian 1988]. Let G be an arbitrary (nonabelian) group given by finitely many generators and relations. Then φ is a construction that successively inserts relations in the middle of words, starting from a word in two elements $u_0, u_1 \in G$. This φ is not likely to have image in co-NP, because of the undecidability of the word problem in group theory [Novikov 1955].

Similarly, in the case of the combinatorial one-way functions proposed below (see §3), Brassard's theorem says nothing. More precisely, if the theorem's hypothesis (1) were to hold in our situation, then it is easy to show that this would imply NP=co-NP.

In other words, in all of these systems an extremely unlikely consequence would immediately follow from the hypothesis of Brassard's theorem. Thus, it is not valid to use Brassard's theorem as an argument against combinatorial cryptography.

Exercises for § 2

1. Find a simple polynomial time certificate that $y \in \mathbb{F}_q^*$ is not in the subgroup generated by g.

2. Let $\mathcal{A}$ be a d-dimensional cellular automaton whose states are $S = \{0, 1\}$ and whose local rule $f : \{0, 1\}^n \to \{0, 1\}$ is addition modulo 2. Show that whether or not $\mathcal{A}$ is reversible depends on the choice of integer n and vectors $\overline{a}_1, \ldots, \overline{a}_n \in \mathbb{Z}^d$ that define the neighborhoods.

§ 3. Concrete Combinatorial–Algebraic Systems

3.1 Polly Cracker

We now describe a general public key cryptosystem, which Fellows has called "Polly Cracker". Let $\mathbb{F}$ be a finite field, and let $T = \{t_i\}_{i=1}^n$ be a set of variables. Alice wants to be able to receive messages $m \in \mathbb{F}$ from Bob. Her secret key is a random vector $y \in \mathbb{F}^n$, and her public key is a set of polynomials $B = \{q_j\}$ in $\mathbb{F}[T]$ such that

$$q_j(y) = 0 \qquad \text{for all } j \ . \tag{2}$$

To send the message m, Bob generates an element

$$p = \sum h_j q_j \tag{3}$$

of the ideal $J \subset \mathbb{F}[T]$ generated by B, and sends her the polynomial

$$c = p + m \ .$$

(Notice that this is *probabilistic* rather than deterministic encryption; see §2.2 of Chapter 1.) When Alice receives the ciphertext polynomial c, she finds m by evaluating it at y:

$$c(y) = p(y) + m = m \ .$$

For example, suppose that $\mathbb{F} = \mathbb{F}_2$, and m is a single bit. The *cracking problem* in the sense of [Selman 1988] (see §5.1 of Chapter 2) for Polly Cracker is then:

INPUT: Generators $B \subset \mathbb{F}_2[T]$ of an ideal J, and a polynomial $c \in \mathbb{F}_2[T]$.
PROMISE: Either $c \in J$ or $c + 1 \in J$.
QUESTION: Is $c \in J$?
TRAPDOOR: A point where J vanishes.

Remark. It is very easy for Alice to construct a pair

$$(\text{private key} = y \ , \quad \text{public key} = B) \ .$$

Namely, she chooses a random y, arbitrary polynomials $\widetilde{q}_j$, and sets $q_j = \widetilde{q}_j - \widetilde{q}_j(y)$. Of course, it is a nontrivial matter for her to choose the keys in such a way that the system is secure.

3.2 Special Cases of Polly Cracker for Famous Combinatorial Problems

We now show how to construct special cases of Polly Cracker for NP-problems such as Graph 3-Coloring and Graph Perfect Code.

Example 3.1. Graph 3-Coloring. (See Example 4.3 of Chapter 2.)
PUBLIC KEY: A graph $G = (V, E)$.
PRIVATE KEY: A proper 3-coloring, i.e., a map $v \mapsto i_v \in \{1, 2, 3\}$ on the vertices $v \in V$ such that $uv \in E \implies i_u \neq i_v$.
A basis $B = B(G)$ of polynomials in the variables $\{t_{v,i} : v \in V, 1 \le i \le 3\}$ is given as follows. Let $B = B_1 \bigcup B_2 \bigcup B_3$, where

$$B_1 = \{t_{v,1} + t_{v,2} + t_{v,3} - 1 : v \in V\} \ ;$$
$$B_2 = \{t_{v,i} t_{v,j} : v \in V, 1 \le i < j \le 3\} \ ;$$
$$B_3 = \{t_{u,i} t_{v,i} : uv \in E, 1 \le i \le 3\} \ .$$

By setting a variable $t_{v,i}$ equal to 1 if the vertex v is colored i and 0 otherwise, we can find a point in the zero set of B if we know the private key (see Exercise 1 below). In Exercise 1 we shall also see that this zero set is non-empty if and only if the graph G is 3-colorable.

Example 3.2. Subset Perfect Code.

PUBLIC KEY: A finite set of variables T and a set of subsets $T_j \subset T$, $j = 1, \ldots, k$, such that $T = \bigcup T_j$.

PRIVATE KEY: A subset $T_0 \subset T$ such that $T_0 \cap T_j$ consists of one element, $j = 1, \ldots, k$.

A basis B of polynomials in the variables $t \in T$ is given as follows. Let $B = B_1 \bigcup B_2$, where

$$B_1 = \left\{ 1 - \sum_{t \in T_j} t \; : \; 1 \leq j \leq k \right\} \; ;$$

$$B_2 = \{ tt' \; : \; t, t' \in T_j, \; t \neq t', \; 1 \leq j \leq k \} \; .$$

Example 3.2a. Graph Perfect Code.*

PUBLIC KEY: A graph $G = (V, E)$.

PRIVATE KEY: A perfect code, i.e., a subset $V' \subset V$ such that every vertex of V is in the neighborhood $N[v]$ of one and only one $v \in V'$. (By definition, $N[v]$ consists of v itself and all vertices joined to v by an edge.)

A basis B of polynomials in the variables $\{ t_v \; : \; v \in V \}$ is given as follows. Let $B = B_1 \bigcup B_2$, where

$$B_1 = \left\{ 1 - \sum_{u \in N[v]} t_u \; : \; v \in V \right\} \; ;$$

$$B_2 = \{ t_u t_{u'} \; : \; u, u' \in N[v], \; u \neq u', \; v \in V \} \; .$$

Another way of describing which products $t_u t_{u'}$ are in B_2 is that they are the ones for which $\text{dist}(u, u') = 1$ or 2, where the distance between two vertices means the minimum number of edges that must be traversed to go from one to the other.

Remark. Example 3.2a is the special case of Example 3.2 where T_j consists of the variables t_u as u ranges over the neighborhood of the j-th vertex of V.

In each of these cases:

• Knowing the graph $G = (V, E)$ or the subsets $T_j \subset T$ (the public key) is equivalent to knowing the basis B for the ideal J.

---

* The term "perfect code" comes not from cryptography but from the theory of error-correcting codes. For example, take the edge-graph of the cube whose vertices are the points (x, y, z) where $x, y, z \in \{0, 1\}$, and note that the two points $(0, 0, 0)$ and $(1, 1, 1)$ form a perfect code. This is a one-error-correcting Hamming code.

• Knowing a solution to the NP-hard combinatorial problem (the private key) is equivalent to knowing a point y at which the ideal J vanishes. (See Exercise 1 below.)

• To send Alice a message m, Bob randomly generates an element of J and adds m to it.

• To decipher the message, Alice simply evaluates this polynomial at y.

This construction is quite general. In fact, one can prove the following

Theorem 3.1. *For any NP search problem one can construct such a system, i.e., a set B of polynomials (of polynomial size) corresponding to an instance of the problem, such that knowing a point at which B vanishes is polynomial time equivalent to knowing a solution to the search problem.*

A complete proof of this theorem has not yet been written down; but see [Fellows and Koblitz 1994b] for a sketch of a proof.

3.3 Generalization of Polly Cracker

Suppose that Alice has a Gröbner basis $G = \{g_1, \ldots, g_l\}$ (see Definition 5.4 of Chapter 3) of an ideal I in $\mathbb{F}[T]$, where T is a set of variables. G is Alice's secret key. Let $S \subset \mathbb{F}[T]$ be the set of all polynomials that cannot be reduced modulo G (see Definition 5.3 of Chapter 3); this is a set of representatives for the quotient ring $\mathbb{F}[T]/I$ (see Exercise 4(c) of Chapter 3, §5). Suppose that the set S is publicly known, even though G is not. A message unit m is an element of S.

Example 3.3. Let $T = \{t_1, \ldots, t_n\}$ and let $y \in \mathbb{F}^n$ be a secret point. Alice sets $G = \{t_1 - y_1, \ldots, t_n - y_n\}$; in this case $S = \mathbb{F}$ is simply the set of constant polynomials.

Next, Alice chooses a set $B = \{q_j\}$ of polynomials in the ideal I. For example, given an arbitrary polynomial $\widetilde{q}_j$, she could set $q_j = \widetilde{q}_j - \overline{q}_j$, where $\overline{q}_j$ is the element of S that she gets from reducing $\widetilde{q}_j$ modulo the Gröbner basis G. Alice's public key is the set of polynomials B. Let J be the ideal generated by these polynomials.

To send a message $m \in S$, Bob randomly chooses an element $p = \sum h_j q_j \in J$ and sends Alice the ciphertext polynomial $c = p + m$. Alice deciphers the message by reducing c modulo the Gröbner basis.

The special case in Example 3.3 when the Gröbner basis is $G = \{t_i - y_i\}$ gives our earlier Polly Cracker system.

Example 3.4. Let $T = \{t_1, \ldots, t_n\}$ and let $y \in \mathbb{F}^n$ be a secret point. Let I be the ideal consisting of polynomials that vanish to total order at least d at the point y; by definition, this is the ideal generated by the set G of monomials of the form $\prod_{i=1}^{n}(t_i - y_i)^{\alpha_i}$ where the α_i are nonnegative integers whose sum is d. It is easy to see that G is the reduced Gröbner basis for I. The set S consists of all polynomials of total degree less than d. The special case $d = 1$ is, of course, Polly Cracker (Example 3.3).

Exercises for § 3

1. (a) In Examples 3.1, 3.2 and 3.2a construct a one-to-one correspondence between private keys and points y at which B vanishes. (b) In each case show that $t^2 - t$ belongs to the ideal J for each variable t.

2. Generalize Example 3.1 to the Graph m-Coloring problem (the search for a proper coloring by m colors, where $m \geq 2$ is an arbitrary fixed integer).

3. Suppose that the field $\mathbb{F}$ contains three cube roots of unity (in particular, its characteristic is not 3). In Example 3.1, instead of the set of variables $\{t_{v,i}\}$ use the set of variables $\{x_v : v \in V\}$. Set $B' = B_1' \bigcup B_2'$, where $B_1' = \{x_v^3 - 1 : v \in V\}$ and $B_2' = \{x_u^2 + x_u x_v + x_v^2 : uv \in E\}$.
(a) Construct a one-to-one correspondence between proper 3-colorings and points at which B' vanishes.
(b) Construct a ring isomorphism between the quotient ring of $\mathbb{F}[\{t_{v,i}\}]$ modulo the ideal J generated by B and the quotient ring of $\mathbb{F}[\{x_v\}]$ modulo the ideal J' generated by B'.

4. Generalize Exercise 3 to m-colorings.

5. Consider the construction in Exercise 3 in the case of the graph consisting merely of two vertices and an edge between them. Show that $B' = \{X^3 - 1, X^2 + XY + Y^2, Y^3 - 1\}$ is already a reduced Gröbner basis for the ideal J'. Let (x_i, y_i), $i = 1, 2, 3, 4, 5, 6$, be the six points corresponding to the proper colorings of the graph (see Exercise 3(a)). Let $f \in \mathbb{F}[X, Y]$. Prove that $f \in J'$ if and only if $f(x_i, y_i) = 0$ for $1 \leq i \leq 6$. In other words, prove that $J' = J''$, where J'' is the ideal of polynomials that vanish at all six points.

6. Prove that Catherine the cryptanalyst can break the cryptosystem in §3.3 if she can find a Gröbner basis $G' = \{g_1', \ldots, g_{l'}'\}$ for the ideal J generated by B. Even though in Chapter 3 we saw that there is an algorithm for finding a Gröbner basis of any ideal, in the present situation its running time is likely to be prohibitively long.

7. Here is a simplified version of the Graph Perfect Code system (Example 3.2a). Let us work over the field $\mathbb{F}_2$, and suppose that Bob wants to send Alice a secret message consisting of a single bit b ("yes" or "no"). He has a copy of Alice's graph (her public key), in which she knows a secret perfect code. Bob randomly assigns a bit to each of the vertices of the graph except for one. He then assigns a bit to the last vertex in such a way that the mod 2 sum of the bits is b. Next, he replaces the bit c_v assigned to each vertex v by a new bit c_v' determined by summing (mod 2) all of the bits that had been assigned to the neighboring vertices: $c_v' = \sum_{u \in N[v]} c_u$. He finally returns the graph to Alice with the bits c_v' annotating the vertices. To decipher the message, Alice takes the sum of c_v' over the perfect code V' (which is her secret key). That is, she has $b = \sum_{v \in V} c_v = \sum_{v \in V'} c_v'$, where the last equality follows from the definition of a perfect code.

(a) Explain how this is a special case of Example 3.2a.

(b) Show how to break the system by linear algebra modulo 2.

(c) Could this function as a Kid Krypto system? (See Definition 7.1 in Chapter 1.)

8. Here is a variant of Exercise 7. We now work over the rational integers $\mathbb{Z}$, and let $m \in \mathbb{Z}$ be a message that Bob wants to send to Alice. He assigns an integer c_v to each vertex v except for one of them, and then assigns an integer to the last vertex in such a way that the sum of all of the integers c_v is m. He next replaces each integer c_v by the integer $c'_v = \sum_{u \in N[v]} c_u$, and returns the graph to Alice along with the integers c'_v. As before, Alice deciphers by summing the c'_v over the perfect code.

(a) Show how to break this system by linear algebra over $\mathbb{Q}$.

(b) Could this Kid Krypto system be of pedagogical value? Could it make high school students eager to learn linear algebra or even to rediscover it by themselves? (See [Koblitz 1997].)

(c) If the graph is r-regular (that is, if every vertex has r edges emanating from it), then show that even without linear algebra it is easy to break the system.

9. Describe one-way constructions of instances of 3-Coloring and of Graph Perfect Code (see Examples 3.1 and 3.2a). That is, the person doing the construction knows a solution, but the instance might seem difficult to someone who sees only the final result and not the process of construction.

10. The Satisfiability problem of symbolic logic was the first problem to be proved to be NP-complete; it is often used as a point of departure in proving results (such as NP-completeness) about other problems. To define the Satisfiability decision problem, we use the symbol p_i for a logical variable, $\neg p_i$ for its negation, and $\vee$ for disjunction (inclusive 'or'). By a *clause* we mean a finite set of p_i or $\neg p_i$ connected by $\vee$, such as $p_1 \vee \neg p_3 \vee p_4$. The input in Satisfiability is a finite set of clauses. The question is whether there exists an assignment of truth values $\{p_i\} \longrightarrow \{T, F\}$ that makes all of the clauses true. Let $\mathbb{F}$ be an arbitrary field.

(a) Show how to construct a special case of Polly Cracker such that the polynomials have a common zero if and only if the corresponding set of clauses is satisfiable. In other words, prove Theorem 3.1 for Satisfiability.

(b) Modify this construction so that there is a one-to-one correspondence between zeros of the polynomial ideal and truth assignments (that is, functions $\{p_i\} \longrightarrow \{T, F\}$) that make all of the clauses true.

11. Show how an adversary (Catherine) can cryptanalyze a Polly Cracker ciphertext c using an *adaptive chosen-ciphertext attack*. What this means is the following. Suppose that two companies B (Bob's company) and C (Cathy's company) are communicating with A (Alice's company) using Alice's public key. On many questions C is cooperating with A, but there is one extremely important customer who is taking competing bids from a group of companies led by A and B and from a different consortium led by C. C knows that B has just sent A the encrypted amount of their bid (suppose that its successive binary digits m_i are each sent as a ciphertext c_i), and she desperately wants to know what it is. So she sends A a

sequence of ciphertexts c_i', supposedly part of a message on an unrelated subject. She then informs A that she had a computer problem, lost her plaintext, and thinks that an incomplete sequence of bits was encrypted for Alice. Could Alice please send her the decrypted bits m_i' that she obtained from the c_i', so that Cathy can reconstruct the correct message and re-encrypt it? Cathy then is able to use the m_i' to find the m_i, because she constructed the c_i' using the c_i and then lied to Alice about it. Alice is willing to give Cathy the m_i' because she is unable to see any connection between the c_i' and the c_i or between the m_i' and the m_i, and because Cathy's request seems reasonable when they are exchanging messages about a matter on which they are cooperating.

§ 4. The Basic Computational Algebra Problem

In Polly Cracker cryptography the underlying computational algebra problem is Ideal Membership:

INPUT: Polynomials $q_j, c \in \mathbb{F}[T]$, where $\mathbb{F}$ is a field and T is a finite set of variables.

QUESTION: Does c belong to the radical of the ideal generated by the q_j? (See Theorem 4.3 of Chapter 3.)

This problem has the following natural *certificates*:

• If "yes", give a natural number N and polynomials $h_j \in \mathbb{F}[T]$ such that $c^N = \sum h_j q_j$.
• If "no", give a point y (with coordinates in an algebraic extension of $\mathbb{F}$) such that $q_j(y) = 0$ for all j but $c(y) \neq 0$.

Because both "yes" and "no" instances have certificates, we might be tempted to conclude that Ideal Membership – like factoring (see Example 4.8 of Chapter 2) – belongs to both NP and co-NP. That would be very wrong, in fact, as wrong as one can possibly be. It can be proved (see Remark 2 below) that Ideal Membership is *neither* in NP nor in co-NP. The difficulty is that in general neither certificate has polynomial size as a function of the input length. On the other hand, the instances of Ideal Membership that arise in our application to cryptography (see §3.1) must have certificates of reasonable size, because Bob will do a limited amount of computation to come up with the h_j and Alice will choose a point y with coordinates in a small field.

Remarks. 1. In the special case $c = 1$, results which give bounds on the degree of h_j or the field extension degree of the coordinates of y are called "effective Nullstellensatz".

2. The Ideal Membership problem is EXPSPACE-hard (see §7.4 of Chapter 2) [Mayr and Meyer 1982]. In particular, this implies that it is in neither NP nor co-NP.

3. In general, the degrees of the h_j grow *doubly exponentially* in the number of variables (see [Möller and Mora 1984] and [Huynh 1986a]). For definitive results in the case of effective Nullstellensatz, see [Kollár 1988].

4. Even when sharply restricted – for example, to the case of only 4 variables or to the case when all of the q_j have the form $T_i - M$ (where T_i is a variable and M is a monomial) – the Ideal Membership problem is NP-hard [Huynh 1986b].

5. It can also be shown that the extension degree of the field generated by the coordinates of y might grow exponentially as a function of the input length.

Exercise for § 4

1. Give an example where the extension degree of the field generated by the coordinates of a point y in a "no" certificate for Ideal Membership grows exponentially or nearly exponentially.

§ 5. Cryptographic Version of Ideal Membership

Mindful of the Merkle–Hellman fiasco (see §1), we must avoid the fallacy of assuming that the intractability of the general Ideal Membership problem implies intractability of the instances that must be solved to break Polly Cracker. Bob does not use h_j of superexponential degree, and Alice does not choose her point y in a field extension of exponentially high degree.

We want cracking the cipher to be difficult not only as a function of the cryptanalyst's input, but also as a function of the work that Bob and Alice have to perform. This motivates the next two definitions.

Definition 5.1. By "phantom input" we mean a string of symbols that is not part of the input but whose length is included in input length. In other words, we suppose that the input includes a string of meaningless symbols of length equal to that of the phantom input.

The cryptographic version of Ideal Membership is a "promise problem", where the promise is that the polynomial in question, namely c, differs from the ideal by a unique element of the message set. In the case of Polly Cracker (§3.1), this means that $p = c - m$ belongs to J for some $m \in \mathbb{F}$ and that J is not the unit ideal. This promise can be certified by giving h_j and y such that (2) and (3) hold.

Definition 5.2. We take the phantom input in a promise problem to be a certificate of correctness of the promise. In the case of Ideal Membership, we call the resulting promise problem Phantom Ideal Membership.

Open Question of Cryptographic Interest. What can be said about the complexity of Phantom Ideal Membership? Could it possibly be polynomial time? (If so, then Polly Cracker is truly cracked.)

§6. Linear Algebra Attacks

There are essentially two ways I know of to attack the cryptosystems in §3. The first method applies to systems of the type in §3.2 that are based on a supposedly hard instance of an NP-hard combinatorial problem. Namely, one tries to solve the underlying combinatorial problem, in the hope that Alice has done a poor job with her one-way construction of an instance of the problem. If one succeeds, then one is in the same position as Alice, and can immediately decrypt any message sent to her.

It is not known whether or not efficient algorithms exist that with a probability close to 100% will produce hard solved instances of an NP-hard problem. In other words, no one has been able to give a systematic way for Alice to carry out a one-way construction of Perfect Code, 3-Coloring, or any other NP-hard problem that has withstood attempts to give a subexponential time algorithm that solves most of the instances constructed. For instance, in 1988 Kučera and Micali thought that they had a method to get hard instances of the NP-complete problem Clique.* However, A. Broder soon found a subexponential time algorithm that solves those instances.

On the other hand, no one has been able to prove that an unexpected consequence (such as P=co-NP) would result from the existence of a polynomial time algorithm to produce hard solved instances. So the matter is wide open.

The second approach to cryptanalysis looks for weaknesses in Bob's construction of the ciphertext c rather than in Alice's construction of the keys. If this approach succeeds, then the cryptanalyst will know a particular secret message m, but will not necessarily be able to decipher the next message that Bob sends to Alice, particularly if he does a better job choosing his coefficient polynomials h_j.

The method is as follows. Suppose that we are in the situation of §3.1 (it is not hard to extend the method to the generalization of Polly Cracker in §3.3). Set

$$\sum h_j q_j = c \qquad \text{up to constant ,}$$

and solve for the unknown h_j. That is, regard the coefficients in the h_j as unknowns, and get linear equations by equating nonconstant monomial terms of $\sum h_j q_j$ and c.

If c and the q_j are "sparse" polynomials – for example, if only $2^{O(d)}$ of their $O(n^d)$ monomial terms are nonzero, where d is the degree and n is the number of variables – then the method in this general form is exponential time. However, a serious attack on Ideal Membership is possible by refining this method, i.e., using "intelligent" linear algebra. The existence of such an attack caused T. Mora and

---

* Given a graph $G = (V, E)$ and an integer k, a k-clique is a subset of k vertices in V all pairs of which are connected by edges in E. The Clique problem asks whether a k-clique exists.

others [1993] to conjecture that Ideal Membership cannot be used to construct a public key system.[*]

Here is one version of the intelligent linear algebra attack. It was proposed by H. W. Lenstra, Jr. (private communication):

"Let C be the set of monomials occurring in c, and let Q_j be the set of monomials occurring in q_j. Then the cryptanalyst might *believe* that any monomial d occurring in h_j is such that $d \cdot Q_j$ intersects C. The set D of those d's is easy to determine and is not too large, and so linear algebra solves the problem in deterministic polynomial time – provided, of course, that the *belief* is correct.

"But to defeat that belief, Bob must artfully build at least one monomial d' into at least one h_j such that d' times *any* term in q_j is canceled in the entire sum (so that it doesn't occur in C). Also, the monomials d' with that property should not be too few and/or too easy to guess, since otherwise the cryptanalyst would simply adjoin those d' to D."

A Polly Cracker cryptosystem is obviously insecure if it succumbs to such a linear algebra attack.

§7. Designing a Secure System

Can a version of Polly Cracker be devised that is secure? (Here we are leaving aside the question of efficiency.) The following is an attempt to design such a system. We shall work with Graph Perfect Code (Example 3.2a). We suppose that

$\mathbb{F} = \mathbb{F}_2$;
the graph $G = (V, E)$ has perfect code V';
$n = \#V$, and $n' = \#V'$;
d =degree of the ciphertext polynomial c.

For convenience, let us also suppose that G is 3-regular (i.e., every vertex has 3 edges emanating from it), in which case $n = 4n'$. Here the order of n and d to have in mind is:

$$n \approx 500 \ , \qquad d \approx 2 \log_2 n \approx 18 \ .$$

---

[*] The authors also cite two theorems to support their skepticism. The first, from [Giusti 1984], states that, even though the degrees of the polynomials in a Gröbner basis can be extremely large, for "almost all" ideals they are not. More precisely, in the parameter space of ideals generated by s polynomials in n variables of degree bounded by D there is a Zariski-open set where the ideals have reduced Gröbner basis consisting of polynomials of degree at most $(n + 1)D - n$. (A "Zariski-open" set is the complement of the zero set of an ideal; see Definition 4.10 of Chapter 3.) The second theorem, from [Dickenstein, Fitchas, Giusti, and Sessa 1991], states that if a function is constructed by adding multiples $h_j q_j$ of elements in an ideal, where the degree of $h_j q_j$ is known to be bounded by D, then in testing Ideal Membership by means of a Gröbner basis one can ignore steps in the algorithm involving polynomials of degree greater than D.

We now describe how to construct a degree-d polynomial c in the variables t_v, $v \in V$, that encrypts $m = 0$ or 1. Recall that the ideal J is generated by all $t_u t_{u'}$ for which $\text{dist}(u, u') = 1$ or 2, and by all $1 - \sum_{u \in N[v]} t_u$ as v ranges over V.

Let d_0 be chosen $\approx d/3$. We construct c_ℓ, $\ell = 1, \ldots, d$, in three stages: (I) $\ell = 1$, (II) $1 < \ell \le d_0$, and (III) $d_0 < \ell \le d$. Then we set $c = c_d$.

Step. I. Construct a linear form c_1 that contains about half of the variables by setting

$$c_1 = \sum_v \sum_{u \in N[v]} t_u ,$$

where the outer sum is taken over a randomly chosen subset of V of cardinality $\approx n'/2$; here the cardinality is even if $m = 0$ and odd if $m = 1$.

Step. II. Let $\mathcal{R}$ denote the following "reduction" modulo J of monomials $\prod t_v^{\alpha_v}$: replace each power $\alpha_v > 1$ by the first power, and replace the monomial by zero if it contains two variables t_u and t_v with $\text{dist}(u, v) = 1$ or 2. Suppose that $c_{\ell-1}$ has been constructed, where $1 < \ell \le d_0$. For each monomial M in $c_{\ell-1}$, select a random vertex v_M, replace M by

$$\mathcal{R}\left(M \sum_{u \in N[v_M]} t_u \right) ,$$

and let c_ℓ denote the resulting sum (after any cancelation).

Step. III. Suppose that $\ell > d_0$, and $c_{\ell-1}$ has been constructed. The construction of c_ℓ is as in Step II, except that the choice of v_M for each M is no longer completely random. Namely, v_M is chosen at a distance 2 from one of the vertices whose variable occurs in M. If the graph is 3-regular, then $\mathcal{R}\left(M \sum_{u \in N[v_M]} t_u \right)$ will consist of at most 2 monomials. Namely, it will consist of $3 - \eta$ monomials, where $\eta \ge 1$ is the number of neighbors of v_M (besides v_M itself) that are at a distance 1 or 2 from some vertex whose variable occurs in M. In particular, if v_M is "surrounded by the vertices in M", then $M \sum_{u \in N[v_M]} t_u$ disappears under the reduction $\mathcal{R}$.

We can visualize the presence of M in the graph as shown in the picture at the bottom of the page. Every vertex whose variable occurs in M is represented

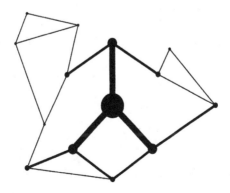

by a HUGE swollen dot. Every vertex at a distance 1 from a huge swollen dot is depicted by a big (but not huge) dot, and every vertex at a distance 2 from a huge swollen dot is depicted by a medium-size dot. All other vertices are small dots. Big thick legs connect the huge dots to the neighboring big dots, and thinner legs connect the big dots to the neighboring medium-size dots.

So we choose the v_M for higher and higher degrees ℓ in such a way that the "spiders" corresponding to the vertices in the monomials M start to "circle their prey". When v_M is surrounded, the corresponding term in c_ℓ vanishes.

So far, this "directed randomness" seems to have thwarted attempted linear algebra attacks. But one cannot have confidence in this approach to constructing a cryptosystem until much more effort has been devoted to investigating such attacks.

We conclude by listing some open questions concerning implementation:

1) What one-way constructions lead to hard instances, say, of 3-Coloring or of Perfect Code?

2) What sparse constructions of polynomials will resist a clever linear algebra attack?

3) Can one get random-self-reducibility (see [Feigenbaum, Kannan, and Nisan 1990]) and hard-on-average (see [Levin 1984]) cracking problems?

4) Can these systems be made efficient?

Chapter 6. Elliptic and Hyperelliptic Cryptosystems

> For there exists a certain Intelligible which you must perceive by the flower of mind.
>   – Beginning of *The Chaldean Oracles** (p. 49 of [Majercik 1989])

Starting in about 1985, the theory of elliptic and hyperelliptic curves over finite fields has been applied to various problems in cryptography: factorization of integers, primality testing, and construction of cryptosystems. In this chapter we shall discuss the last of these. One of the main reasons for interest in cryptosystems based on elliptic and hyperelliptic curves is that these curves are a source of a tremendous number of finite abelian groups having a rich algebraic structure.

In many ways the elliptic curve groups and the jacobian groups of hyperelliptic curves are analogous to the multiplicative group of a finite field. However, they have two advantages: there are far more of them, and they seem to provide the same security with smaller key size. We shall be more specific about this later.

We shall start by giving the basic definitions and facts about elliptic curves. Our account will emphasize concrete examples and algorithms rather than proofs and the general theory. For a more systematic treatment of elliptic curves, see [Silverman 1986], [Husemöller 1987], and [Koblitz 1993].

After that we shall describe some cryptosystems based on elliptic curves and briefly discuss some open questions that arise from cryptographic applications. In §§5–6 we shall treat hyperelliptic curves and cryptosystems.

§ 1. Elliptic Curves

1.1 The Equation

An elliptic curve E over a field $\mathbb{F}$ is a curve that is given by an equation of the form

$$Y^2 + a_1 XY + a_3 Y = X^3 + a_2 X^2 + a_4 X + a_6 , \qquad a_i \in \mathbb{F} . \qquad (1)$$

We let $E(\mathbb{F})$ denote the set of points $(x, y) \in \mathbb{F}^2$ that satisfy this equation, along with a "point at infinity" denoted O. If $\mathbb{K}$ is any extension field of $\mathbb{F}$, then $E(\mathbb{K})$ denotes the set of $(x, y) \in \mathbb{K}^2$ that satisfy (1), along with O. In order for the curve (1) to be an *elliptic curve* it must be smooth. This means that there is no point of $E(\overline{\mathbb{F}})$ (recall that $\overline{\mathbb{F}}$ denotes the algebraic closure of $\mathbb{F}$) where both

---

* My thanks to Ron Rivest for piquing my interest in Chaldean poetry.

partial derivatives vanish (see Definition 1.6 of Chapter 3). In other words, the two equations

$$a_1 Y = 3X^2 + 2a_2 X + a_4 \ , \qquad 2Y + a_1 X + a_3 = 0 \tag{2}$$

cannot be simultaneously satisfied by any $(x, y) \in E(\overline{\mathbb{F}})$.

If $\mathbb{F}$ is not of characteristic 2, then without loss of generality we may suppose that $a_1 = a_3 = 0$ (see Exercise 1(a) below). In the important case of characteristic 2 we have the so-called "supersingular" case with $Y^2 + a_3 Y$ on the left in (1) and the "nonsupersingular" case with $Y^2 + a_1 XY$ on the left; in the latter case without loss of generality we may suppose that $a_1 = 1$ (see Exercise 1(b) below). (In characteristic 2 we may also suppose that $a_2 = 0$ in the supersingular case and that $a_4 = 0$ in the nonsupersingular case; see Exercise 3(b) below.) The reason for the subscripts in a_1 and a_3 on the left of (1) and in a_2, a_4, and a_6 on the right will be explained soon.

If the characteristic of $\mathbb{F}$ is neither 2 nor 3, then, after simplifying the left side of (2), by a linear change of variables (namely, $X \to X - \frac{1}{3}a_2$) we can also remove the X^2-term. That is, without loss of generality we may suppose that our elliptic curve is given by an equation of the form

$$Y^2 = X^3 + aX + b \ , \qquad a, b \in \mathbb{F} \ , \qquad \operatorname{char} \mathbb{F} \neq 2, 3 \ . \tag{3}$$

In this case the condition that the curve be smooth is equivalent to requiring that the cubic on the right have no multiple roots. This holds if and only if the *discriminant* of $X^3 + aX + b$, which is $-(4a^3 + 27b^2)$, is nonzero. (Recall that the discriminant of a monic polynomial of degree d with roots $r_1, \ldots, r_d$ is $\prod_{i \neq j}(r_i - r_j) = (-1)^{d(d-1)/2} \prod_{i<j}(r_i - r_j)^2$.)

For any extension field $\mathbb{K}$ of $\mathbb{F}$, the set $E(\mathbb{K})$ forms an abelian group whose identity element is O. To explain the rules for adding points, it is best to look first at elliptic curves defined over the real number field $\mathbb{R}$. For example, the graph of the elliptic curve $Y^2 = X^3 - X$ is shown on the next page.

Notice that for large X the curve goes out to infinity much like the function $Y = X^{3/2}$, which can be parameterized by setting $X = T^2$ and $Y = T^3$. We often say that "X has degree 2" and "Y has degree 3". The subscripts of the a's in (1) indicate the degrees that must be given to the coefficients in order that the equation (1) be homogeneous, that is, in order that each term have total degree 6. That is the reason why it is traditional to label the subscripts in (1) in a way that at first looks peculiar.

1.2 Addition Law

Definition 1.1. Let E be an elliptic curve over the real numbers given by equation (3), and let P and Q be two points on E. We define the negative of P and the sum $P + Q$ according to the following rules:

1) If P is the point at infinity O, then we define $-P$ to be O. For any point Q we define $O + Q$ to be Q; that is, O serves as the additive identity ("zero element")

of the group of points. In what follows, we shall suppose that neither P nor Q is the point at infinity.

2) The negative $-P$ is the point with the same x-coordinate as P but negative y-coordinate; that is, $-(x,y) = (x,-y)$. It is obvious from equation (3) that $(x,-y)$ is on the curve whenever (x,y) is. If $Q = -P$, then we define $P + Q$ to be the point at infinity O.

3) If P and Q have different x-coordinates, then we shall soon show that the line $\ell = \overline{PQ}$ intersects the curve in exactly one more point R (unless ℓ is tangent to the curve at P, in which case we take $R = P$, or at Q, in which case we take $R = Q$). Then we define $P + Q$ to be $-R$, that is, the mirror image (with respect to the x-axis) of the third point of intersection. The geometrical construction that gives $P + Q$ is illustrated in the drawing below.

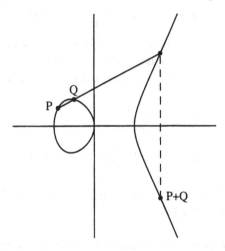

4) The final possibility is that $P = Q$. Then let ℓ be the tangent line to the curve at P, let R be the only other point of intersection of ℓ with the curve, and define $2P = -R$. (R is taken to be P if the tangent line has a "double tangency" at P, in other words, if P is a point of inflection.)

The above set of rules can be summarized in the following succinct manner:

the sum of the three points where a line intersects the curve is zero.

If the line passes through the point at infinity O, then this relation has the form $P + \widetilde{P} + O = O$ (where P and $\widetilde{P}$ are symmetrical points), i.e., $\widetilde{P} = -P$. Otherwise, it has the form $P + Q + R = O$, where P, Q, and R are the three points in rule 3) or 4).

We now show why there is exactly one more point where the line ℓ through P and Q intersects the curve; at the same time we will derive a formula for the coordinates of this third point, and hence for the coordinates of $P + Q$.

Let (x_1, y_1), (x_2, y_2) and (x_3, y_3) denote the coordinates of P, Q and $P + Q$, respectively. We want to express x_3 and y_3 in terms of x_1, y_1, x_2, y_2. Suppose that we are in case 3) in the definition of $P + Q$, and let $y = \alpha x + \beta$ be the equation of the line through P and Q (which is not a vertical line in case 3)). Then $\alpha = (y_2 - y_1)/(x_2 - x_1)$, and $\beta = y_1 - \alpha x_1$. A point $(x, \alpha x + \beta) \in \ell$ lies on the elliptic curve if and only if $(\alpha x + \beta)^2 = x^3 + ax + b$. Thus, there is one intersection point for each root of the cubic equation $x^3 - (\alpha x + \beta)^2 + ax + b$. We already know that there are the two roots x_1 and x_2, because $(x_1, \alpha x_1 + \beta)$, $(x_2, \alpha x_2 + \beta)$ are the points P, Q on the curve. Since the sum of the roots of a monic polynomial is equal to minus the coefficient of the second-to-highest power, we conclude that the third root in this case is $x_3 = \alpha^2 - x_1 - x_2$. This leads to an expression for x_3, and hence for both coordinates of $P + Q = (x_3, -(\alpha x_3 + \beta))$, in terms of x_1, x_2, y_1, y_2:

$$x_3 = \left(\frac{y_2 - y_1}{x_2 - x_1} \right)^2 - x_1 - x_2 \; ;$$

$$y_3 = -y_1 + \left(\frac{y_2 - y_1}{x_2 - x_1} \right)(x_1 - x_3) \; .$$

(4)

The case when $P = Q$ is similar, except that α is now the derivative dy/dx at P. Implicit differentiation of equation (3) leads to the formula $\alpha = (3x_1^2 + a)/2y_1$, and so we obtain the following formulas for the coordinates of twice P:

$$x_3 = \left(\frac{3x_1^2 + a}{2y_1} \right)^2 - 2x_1 \; ;$$

$$y_3 = -y_1 + \left(\frac{3x_1^2 + a}{2y_1} \right)(x_1 - x_3) \; .$$

(5)

Example 1.1. Let $P = (0, 0)$ on the elliptic curve $Y^2 + Y = X^3 - X^2$. Find $2P = P + P$ and $3P = P + 2P$.

Solution. We first transform the equation to the form (3) by making the change of variables $Y \to Y - \frac{1}{2}$, $X \to X + \frac{1}{3}$. On this curve P becomes $Q = (-\frac{1}{3}, \frac{1}{2})$. Using (5), we obtain $2Q = (\frac{2}{3}, -\frac{1}{2})$. Then from (4) we have $3Q = 2Q + Q = (\frac{2}{3}, \frac{1}{2})$. Notice that $3Q = -(2Q)$, and hence Q is a point of order 5, i.e., $5Q = O$. Back on the original curve we have $2P = (1, -1)$, $3P = (1, 0) = -2P$.

There are several ways of proving that the above definition of $P + Q$ makes the points on an elliptic curve into an abelian group. One can use an argument from projective geometry, a complex analytic argument with doubly periodic functions, or an algebraic argument involving divisors on curves. The only group law that is not an immediate consequence of the geometrical rules 1)–4) is the associative law. That can be proved from the following fact from the projective geometry of cubic curves (see Exercise 4 below):

Proposition. *Let l_1, l_2, l_3 be three lines that intersect a cubic in nine points $P_1, \ldots, P_9$ (counting multiplicity), and let l'_1, l'_2, l'_3 be three lines that intersect the cubic in nine points $Q_1, \ldots, Q_9$. If $P_i = Q_i$ for $i = 1, \ldots, 8$, then also $P_9 = Q_9$.*

As in any abelian group, we use the notation nP to denote P added to itself n times if n is positive, and $-P$ added to itself $|n|$ times if n is negative.

We have not yet said much about the "point at infinity" O. By definition, it is the identity of the group law. In the above graph of the curve $Y^2 = X^3 - X$, this point should be visualized as sitting infinitely far up the y-axis, in the limiting direction of the ever-steeper tangents to the curve. It is the "third point of intersection" of any vertical line with the curve; that is, such a line has points of intersection of the form (x_1, y_1), $(x_1, -y_1)$ and O. A more natural way to introduce the point O is as follows.

1.3 Projective Coordinates

By the *projective plane* over the field $\mathbb{F}$ we mean the set of equivalence classes of triples (X, Y, Z) (not all components zero) where two triples are said to be equivalent if they are a scalar multiple of one another; in other words, $(X', Y', Z') \sim (X, Y, Z)$ if $(\lambda X', \lambda Y', \lambda Z') = (X, Y, Z)$ for some $\lambda \in \mathbb{F}$. Such an equivalence class is called a *projective point*. If a projective point has nonzero Z, then there is one and only one triple in its equivalence class of the form $(x, y, 1)$: simply set $x = X/Z$, $y = Y/Z$. Thus, the projective plane can be identified with all points (x, y) of the ordinary ("affine") plane plus the points for which $Z = 0$. The latter points make up what is called the *line at infinity*; roughly speaking, it can be visualized as the "horizon" on the plane. Any equation $F(X, Y) = 0$ of a curve in the affine plane corresponds to a *homogeneous* equation $\widetilde{F}(X, Y, Z) = 0$ satisfied by the corresponding projective points: simply replace X by X/Z and Y by Y/Z and multiply by a power of Z to clear the denominators. For example, if we apply this procedure to the affine equation (3) of an elliptic curve, we obtain its "projective equation" $Y^2 Z = X^3 + aXZ^2 + bZ^3$. The latter equation is satisfied by a projective point (X, Y, Z) with $Z \neq 0$ if and only if the corresponding affine point (x, y), where $x = X/Z$ and $y = Y/Z$, satisfies (3). In addition to the points with $Z \neq 0$, what projective points (X, Y, Z) satisfy the equation $\widetilde{F} = 0$? Setting $Z = 0$ in the equation, we obtain $0 = X^3$, which means that $X = 0$. But the only equivalence class of triples (X, Y, Z) with both X and Z zero is the class of $(0, 1, 0)$. This is the point we call O. It is the point on the intersection of the y-axis with the line at infinity.

1.4 Elliptic Curves over $\mathbb{C}$

We saw that if $\mathbb{F}$ is any field of characteristic $\neq 2, 3$, then the equation of an elliptic curve can be given in the form (3). The algebraic formulas (4)–(5) for adding points on an elliptic curve over the reals actually make sense over any such field $\mathbb{F}$, and can be shown to give an abelian group law on the curve.

In particular, let E be an elliptic curve defined over the field $\mathbb{C}$ of complex numbers. Thus, E is the set of pairs (x, y) of complex numbers satisfying equation (3), together with the point at infinity O. Although E is a "curve", if we think in terms of familiar geometrical pictures, it is 2-dimensional. That is, it is a surface in the 4-real-dimensional space whose coordinates are the real and imaginary parts of x and y. We now describe how E can be visualized as a surface.

Let L be a *lattice* in the complex plane. This means that L is the abelian group of all integer combinations of two complex numbers ω_1 and ω_2 (where ω_1 and ω_2 span the plane; that is, they do not lie on the same line through the origin). We write $L = \mathbb{Z}\omega_1 + \mathbb{Z}\omega_2$. For example, if $\omega_1 = 1$ and $\omega_2 = i$, then L is the Gaussian integers, the square grid consisting of all complex numbers with integer real and imaginary parts.

Given an elliptic curve (3) over the complex numbers, it turns out that there exist a lattice L and a complex function, called the "Weierstrass $\wp$-function" and denoted $\wp_L(z)$, that has the following properties:

(1) $\wp_L(z)$ is analytic except for a double pole at each point of L;

(2) $\wp_L(z)$ satisfies the differential equation $\wp_L'^2 = \wp_L^3 + a\wp_L + b$, and hence for any $z \notin L$ the point $(\wp_L(z), \wp_L'(z))$ lies on the elliptic curve E;

(3) two complex numbers z_1 and z_2 give the same point $(\wp_L(z), \wp_L'(z))$ on E if and only if $z_1 - z_2 \in L$;

(4) the map that associates any $z \notin L$ to the corresponding point $(\wp_L(z), \wp_L'(z))$ on E and associates any $z \in L$ to the point at infinity $O \in E$ gives a 1-to-1 correspondence between E and the quotient of the complex plane by the subgroup L (denoted $\mathbb{C}/L$);

(5) this 1-to-1 correspondence is an isomorphism of abelian groups. In other words, if z_1 corresponds to the point $P \in E$ and z_2 corresponds to $Q \in E$, then the complex number $z_1 + z_2$ corresponds to the point $P + Q$.

Thus, we can think of the abelian group E as equivalent to the complex plane modulo a suitable lattice. To visualize the latter group, note that every equivalence class $z + L$ has one and only one representative in the "fundamental parallelogram" consisting of complex numbers of the form $u\omega_1 + v\omega_2$, $0 \leq u, v < 1$ (for example, if L is the Gaussian integers, the fundamental parallelogram is the unit square). Since opposite points on the parallel sides of the boundary of the parallelogram differ by a lattice point, they are equal in $\mathbb{C}/L$. That is, we think of them as "glued together". If we visualize this – folding over one side of the parallelogram to meet the opposite side (obtaining a segment of a cylinder) and then folding over again and gluing the opposite circles – we see that we obtain a "torus" (surface of a donut), pictured below.

As a group, the torus is the product of two copies of a circle. That is, its points can be parameterized by ordered pairs of angles (α, β). (More precisely, if

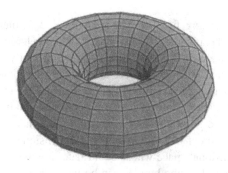

the torus was obtained from the lattice $L = \mathbb{Z}\omega_1 + \mathbb{Z}\omega_2$, then we write an element in $\mathbb{C}/L$ in the form $u\omega_1 + v\omega_2$ and take $\alpha = 2\pi u$, $\beta = 2\pi v$.) Thus, we can think of an elliptic curve over the complex numbers as a generalization to two real dimensions of the circle in the real plane. In fact, this analogy goes much further than one might think. The "elliptic functions" (which tell us how to go back from a point $(x, y) \in E$ to the complex number z for which $(x, y) = (\wp_L(z), \wp'_L(z))$) turn out to have some properties analogous to the familiar arcsine function (which tells us how to go back from a point (x, y) on the unit circle to the real number α that corresponds to that point when we "wrap" the real number line around the circle). In the algebraic number theory of elliptic curves, one finds a deep analogy between the coordinates of the "n-division points" on an elliptic curve (the points P such that nP is the identity O) and the n-division points on the unit circle (which are the n-th roots of unity in the complex plane).

The *order* of a point P on an elliptic curve is the smallest integer n such that $nP = O$; of course, such a finite n need not exist. It is often of interest to find points P of finite order on an elliptic curve, especially for elliptic curves defined over $\mathbb{Q}$.

Example 1.2. Find the order of $P = (2, 3)$ on $y^2 = x^3 + 1$.

Solution. Using (5), we find that $2P = (0, 1)$ and $4P = 2(2P) = (0, -1)$. Thus, $4P = -2P$, and so $6P = O$. Hence, the order of P is 2, 3 or 6. But $2P = (0, 1) \neq O$, and if P had order 3, then we would have $4P = P$, which is not true. So P has order 6.

An important concept in the study of elliptic curves is *complex multiplication*. We say that a curve E defined over $\mathbb{C}$ has complex multiplication by a complex number $\alpha \notin \mathbb{Z}$ if multiplication by α maps the lattice L to itself. It is not hard to show that if E has complex multiplication – that is, if such an α exists – then $\mathbb{Q}(\alpha)$ is an imaginary quadratic field (called the *CM-field* of E). In that case the set of all α such that $\alpha L \subset L$ forms a subring of finite index (called an "order") in the ring of integers of this imaginary quadratic field.

Whenever $\alpha L \subset L$, multiplication by α corresponds to an *endomorphism* of the elliptic curve E. For instance, if E is the elliptic curve in Example 1.2, it turns out that L is a grid of vertices of equilateral triangles in the plane. That is, up to scaling it is the ring $\mathbb{Z}[\zeta]$, where $\zeta = (-1 + \sqrt{-3})/2$. This ring $\mathbb{Z}[\zeta]$ is also the set of α such that $\alpha L \subset L$. The case $\alpha = \zeta$ corresponds to the automorphism of the elliptic curve given by taking $P = (x, y)$ to the point $(\zeta x, y)$.

Similarly, the elliptic curve $Y^2 = X^3 - X$ has complex multiplication by the Gaussian integer ring $\mathbb{Z}[i]$. The case $\alpha = i$ corresponds to the automorphism $(x, y) \mapsto (-x, iy)$.

1.5 Elliptic Curves over $\mathbb{Q}$

In equation (3), if a and b are rational numbers, it is natural to look for rational solutions (x, y). There is a vast theory of elliptic curves over the rationals. Mordell [1922] proved that the abelian group is finitely generated. This means that it consists of a finite "torsion subgroup" E_{tors}, consisting of the rational points of finite order, plus the subgroup generated by a finite number of points of infinite order:

$$E(\mathbb{Q}) \approx E_{\text{tors}} \oplus \mathbb{Z}^r .$$

The number r of generators needed for the infinite part is called the *rank*; it is zero if and only if the entire group of rational points is finite. The study of the rank r and other features of the group of points on an elliptic curve over $\mathbb{Q}$ is related to many interesting questions in number theory and algebraic geometry. We shall discuss this further in §§3 and 4.

1.6 Characteristics 2 and 3

If $\text{char}(\mathbb{F}) = 2$, then an elliptic curve cannot be put in the form (3); in fact, the curve (3) is never smooth in characteristic 2 (see Exercise 2 below). In the case of characteristic 3, one cannot eliminate the $a_2 X^2$-term if it is not already zero. Thus, we cannot use the formulas (4)–(5) directly.

However, we can find formulas analogous to (4)–(5) that apply to elliptic curves whose equation has the more general form (1), which can be used in any characteristic. Again we first suppose that our elliptic curve is defined over $\mathbb{R}$, and we translate the geometrical addition rules 1)–4) into equations for the x- and y-coordinates of $P + Q$ and $2P$. The resulting formulas are aesthetically rather unappealing, and will not be given here. What we do need are the formulas analogous to (4) and (5) that one gets

1) when $a_1 = a_3 = 0$ in (1) but a_2 is not necessarily zero, so that we can work in characteristic 3:

$$x_3 = \left(\frac{y_2 - y_1}{x_2 - x_1}\right)^2 - a_2 - x_1 - x_2 , \qquad y_3 = -y_1 + \left(\frac{y_2 - y_1}{x_2 - x_1}\right)(x_1 - x_3) \quad (6)$$

when adding distinct points; and

$$x_3 = \left(\frac{3x_1^2 + 2a_2 x_1 + a_4}{2y_1}\right)^2 - a_2 - 2x_1 ,$$

$$y_3 = -y_1 + \left(\frac{3x_1^2 + 2a_2 x_1 + a_4}{2y_1}\right)(x_1 - x_3) \tag{7}$$

when doubling a point (note that in characteristic 3 the slope term here simplifies to $(a_2 x_1 - a_4)/y_1$);

2) when $a_3 = a_4 = 0$ in (1) but a_1 is nonzero and may be assumed to be equal to 1 (see Exercises 1(b) and 3(b) below), and char($\mathbb{F}$) = 2 (the nonsupersingular case):

$$x_3 = \left(\frac{y_1 + y_2}{x_1 + x_2}\right)^2 + \frac{y_1 + y_2}{x_1 + x_2} + x_1 + x_2 + a_2 ,$$

$$y_3 = \left(\frac{y_1 + y_2}{x_1 + x_2}\right)(x_1 + x_3) + x_3 + y_1 \tag{8}$$

when adding distinct points; and

$$x_3 = x_1^2 + \frac{a_6}{x_1^2} , \qquad y_3 = x_1^2 + \left(x_1 + \frac{y_1}{x_1}\right)x_3 + x_3 \tag{9}$$

when doubling a point;

3) when $a_1 = a_2 = 0$ in (1) but $a_3 \neq 0$, and char($\mathbb{F}$) = 2 (the supersingular case):

$$x_3 = \left(\frac{y_1 + y_2}{x_1 + x_2}\right)^2 + x_1 + x_2 , \qquad y_3 = \left(\frac{y_1 + y_2}{x_1 + x_2}\right)(x_1 + x_3) + y_1 + a_3 \tag{10}$$

when adding distinct points; and

$$x_3 = \frac{x_1^4 + a_4^2}{a_3^2} , \qquad y_3 = \left(\frac{x_1^2 + a_4}{a_3}\right)(x_1 + x_3) + y_1 + a_3 \tag{11}$$

when doubling a point.

In all cases the rules for adding a point $P_1 = (x_1, y_1)$ to $P_2 = (x_2, y_2)$ to obtain $P_3 = P_1 + P_2 = (x_3, y_3)$ can be written in the form of rational expressions for x_3 and y_3 in terms of x_1, x_2, y_1, y_2 and the coefficients a_i.

Note that for an elliptic curve in the general form (1) the negative of a point $P = (x, y)$ is the point $-P = (x, -a_1 x - a_3 - y)$.

1.7 Elliptic Curves over a Finite Field

For the rest of this section we shall let $\mathbb{F}$ be the finite field $\mathbb{F}_q$ of $q = p^f$ elements. Let E be an elliptic curve defined over $\mathbb{F}_q$. If $p \neq 2, 3$, then we suppose that E is given by an equation of the form (3). If $p = 3$, then we also need to allow an X^2-term on the right in (3). If $p = 2$, then there are two cases: the nonsupersingular case

$$Y^2 + XY = X^3 + a_2 X^2 + a_6 \tag{12}$$

and the supersingular case

$$Y^2 + a_3 Y = X^3 + a_4 X + a_6 \tag{13}$$

(see Exercises 1(b) and 3(b) below).

If an elliptic curve E is defined over $\mathbb{F}_q$, then it is also defined over $\mathbb{F}_{q^r}$ for $r = 1, 2, \ldots$, and so it is meaningful to look at solutions – called "$\mathbb{F}_{q^r}$-points" – in extension fields of the defining equation of the curve. We let N_r denote the number of $\mathbb{F}_{q^r}$-points on E. (Thus, $N_1 = N$ is the number of points with coordinates in our "ground field" $\mathbb{F}_q$.)

From the numbers N_r one forms the "generating series" $Z(E/\mathbb{F}_q; T)$, which is the formal power series defined by setting

$$Z(E/\mathbb{F}_q; T) = e^{\sum N_r T^r / r}, \tag{14}$$

in which T is an indeterminate, the notation $E/\mathbb{F}_q$ designates the elliptic curve and the field we're taking as our ground field, and the sum on the right is over all $r = 1, 2, \ldots$. It can be shown that the series obtained by taking the infinite product of the exponential power series $e^{N_r T^r / r}$ actually has positive integer coefficients. This power series is called the *zeta-function* of the elliptic curve (over $\mathbb{F}_q$), and is a very important object associated with E.

The following theorem of Hasse, which says that this zeta-function has a rather simple form, is of great practical value, since it shows how to determine all of the N_r once one knows N_1.

Theorem 1.1. *The zeta-function of an elliptic curve is a rational function of T having the form*

$$Z(E/\mathbb{F}_q; T) = \frac{1 - aT + qT^2}{(1 - T)(1 - qT)}, \tag{15}$$

where only the coefficient of T in the numerator depends on the particular elliptic curve E. This coefficient is related to $N = N_1$ as follows: $N = q+1-a$. In addition, the discriminant of the quadratic polynomial in the numerator is negative or zero (that is, $a^2 \leq 4q$) and so this polynomial has two complex conjugate roots α, $\overline{\alpha}$ both of absolute value $\sqrt{q}$. (More precisely, $1/\alpha$ and $1/\overline{\alpha}$ are the roots, and α, $\overline{\alpha}$ are the "reciprocal roots".)

For a proof, see § V.2 of [Silverman 1986].

Corollary 1.1. *Let N_r denote the number of $\mathbb{F}_{q^r}$-points on E, and set $N = N_1$ and $a = q + 1 - N$. Let α and $\overline{\alpha}$ be the roots of the quadratic polynomial $T^2 - aT + q$. Then*

$$N_r = |\alpha^r - 1|^2 = q^r + 1 - \alpha^r - \overline{\alpha}^r, \tag{16}$$

where $|\ \ |$ denotes the usual complex absolute value.

The corollary is an immediate consequence of Theorem 1.1, as we see by writing the numerator of the right side of (15) in the form $(1 - \alpha T)(1 - \overline{\alpha} T)$, replacing the left side by (14), and then taking the logarithm of both sides. Here we use the identity $\ln(1 - cT) = -\sum c^r T^r / r$. This corollary shows how to determine all of the N_r once you know $N = N_1$.

Example 1.3. The zeta-function of the elliptic curve $Y^2 + Y = X^3$ over $\mathbb{F}_2$ is easily computed from the fact that there are three $\mathbb{F}_2$-points. (Always remember to include the point at infinity when counting the number of points on an elliptic curve.) We find that

$$Z(E/\mathbb{F}_2;T) = (1 + 2T^2)/(1 - T)(1 - 2T) \ .$$

Thus, the reciprocal roots of the numerator are $\pm i\sqrt{2}$. This leads to the formula

$$N_r = \begin{cases} 2^r + 1 \ , & \text{if } r \text{ is odd} \ ; \\ 2^r + 1 - 2(-2)^{r/2} \ , & \text{if } r \text{ is even} \ . \end{cases} \tag{17}$$

Remark. In general, an elliptic curve is said to be *supersingular* if p divides the coefficient a in (15), or equivalently, if $N \equiv 1 \pmod{p}$. In Exercise 17 below, we see that in the case $p = 2$ our earlier use of the terms "supersingular" and "nonsupersingular" agrees with this definition. The equation in Example 1.3, considered over $\mathbb{F}_q$, gives a supersingular elliptic curve whenever $q \equiv 2 \pmod 3$ (see Exercise 10 below).

Another corollary of Theorem 1.1 is the following basic fact (which often goes by the name of "Hasse's theorem").

Corollary 1.2. *The number N of $\mathbb{F}_q$-points on an elliptic curve defined over $\mathbb{F}_q$ lies in the interval*

$$q + 1 - 2\sqrt{q} \le N \le q + 1 + 2\sqrt{q} \ . \tag{18}$$

This corollary follows because $|q + 1 - N| = |a| \le \sqrt{4q}$ by Theorem 1.1.

Thus, Hasse's theorem says that the size of the group E over $\mathbb{F}_q$ is fairly close to that of the finite field itself. Starting with the seminal paper [Schoof 1985], a great deal of effort has been put into developing efficient algorithms to determine $N = \#E$ for an arbitrary elliptic curve E. After work by A. O. L. Atkin, N. Elkies, F. Morain, J. Miller, J. Buchmann and his students, and S. A. Vanstone and his students, it has become practical to compute $\#E$ over fields $\mathbb{F}_q$ where q has several hundred digits. For example, in [Lercier and Morain 1995 and 1996] $\#E$ is computed for an arbitrary elliptic curve over $\mathbb{F}_q$ where $q = p = 10^{499} + 153$ and where $q = 2^{1301}$. See also [Lehmann, Maurer, Müller, and Shoup 1994].

On the other hand, it is important to note that there are many elliptic curves for which $\#E$ can be easily computed without using the sophisticated ideas of Schoof, Atkin, and others.

Example 1.4. Let E be the elliptic curve $Y^2 = X^3 - X$ defined over $\mathbb{F}_p$. In Exercise 10 below we shall consider the supersingular case when $p \equiv 3 \pmod 4$. Here we suppose that we are in the more interesting nonsupersingular case when $p \equiv 1 \pmod 4$. It can be shown (see §2 of Chapter 2 in [Koblitz 1993] or §4 of Chapter 18 in [Ireland and Rosen 1990]) that the number of points $N = p + 1 - a$ on E can be found by writing p as the sum of two squares $p = c^2 + d^2$ and setting $a = 2c$, where c is uniquely determined by requiring that the following congruence of Gaussian integers hold: $c + id \equiv 1 \pmod{2 + 2i}$. This means that c is odd and d is even; and we replace c by $-c$ if either $4|d$ and $c \equiv 3 \pmod 4$ or else $4 \nmid d$ and $c \equiv 1 \pmod 4$. Here are the first few cases: $p = 5 = (-1)^2 + 2^2$, $N = 5 + 1 - 2(-1) = 8$; $p = 13 = 3^2 + 2^2$, $N = 13 + 1 - 2(3) = 8$; $p = 17 = 1^2 + 4^2$, $N = 17 + 1 - 2(1) = 16$; and so on.

Even if $p \equiv 1 \pmod 4$ is an extremely large prime, it is not hard to write it as a sum of squares. To do this, one first finds a square root y of -1 in $\mathbb{F}_p$ (if we choose a random $x \in \mathbb{F}_p^*$ and raise it to the $\left(\frac{p-1}{4}\right)$-th power, there is a 50% chance that we'll get ± 1 and a 50% chance that we'll get a square root of -1). One then uses the Euclidean algorithm in the Gaussian integer ring to find the greatest common divisor of the Gaussian integers p and $y + i$. This g.c.d. will coincide with $c + di$ or $c - di$ up to a factor of ± 1 or $\pm i$. See Exercise 14 in §2 of Chapter 3.

1.8 Square Roots

We now give a probabilistic algorithm for finding points on an elliptic curve E defined over a finite field $\mathbb{F}_q$. We shall suppose that q is odd; when q is a power of 2, one can use the method in Exercises 3–4 of §2.

Since q is odd, we may suppose that the equation of E is in the form $Y^2 = f(X)$, where $f(X)$ is a cubic polynomial. If we choose values of x at random, then every time $f(x)$ is a square in $\mathbb{F}_q^*$ there are two points of the form $(x, \pm y)$ on the curve. From Hasse's theorem (Corollary 1.2) we see that this happens about 50% of the time. For each x that is not a root of $f(x)$, it is easy to determine whether or not $f(x)$ is a square – simply compute $f(x)^{(q-1)/2} = \pm 1$; $f(x)$ is a square if and only if you get $+1$. In what follows we suppose that we have found a value of x such that $z = f(x)$ is a square in $\mathbb{F}_q^*$. It remains to find a square root $y \in \mathbb{F}_q^*$ of z.

If we happen to have $q \equiv 3 \pmod 4$, then all we have to do is take $y = z^{(q+1)/4}$, at which point $y^2 = z \cdot z^{(q-1)/2} = z$. If $q \equiv 1 \pmod 4$, then we use an "approximation" procedure due to Shanks [1972] that depends on the highest power of 2 dividing $q - 1$. Suppose that $q - 1 = 2^s t$, where t is odd and $s \geq 2$. Let u be a non-square in $\mathbb{F}_q^*$; such an element can be found by choosing u at random until you find one for which $u^{(q-1)/2} = -1$. Set $v = u^t$; then v is a primitive (2^s)-th root of unity in $\mathbb{F}_q$. We get an "approximate" solution y_1 to the equation $y^2 = z$ by setting $y_1 = z^{(t+1)/2}$. Then $y_1^2 = z \cdot z^t$. Since z^t is a (2^{s-1})-th root of unity, there is a power v^l such that $y = y_1 v^{-l}$ is a square root of z. Equivalently, we need

$$v^{2l} = y_1^2/z = z^t . \tag{19}$$

The (2^s)-th root of unity v^l is the "correction term" that we need to convert y_1 to a square root of z. We find the binary digits in $l = l_0 + l_1 \cdot 2 + l_2 \cdot 2^2 + \cdots + l_{s-2} 2^{s-2}$ inductively, starting with l_0. Raising both sides of (19) to the (2^{s-2})-th power, we see that $l_0 = 0$ if and only if we obtain 1 on the right; otherwise $l_0 = 1$. We next raise both sides of (19) to the (2^{s-3})-th power to determine whether l_1 is 0 or 1. We continue in this way until we finally determine $l_0, \ldots, l_{s-2}$ so that (19) holds. This completes our description of the algorithm.

It is easy to verify that the above probabilistic algorithm for finding a point on E takes time $O(\ln^3 q)$ (see Theorem 2.7 in Chapter 3).

One might ask whether there is a *deterministic* polynomial-time algorithm for finding points (other than the point at infinity) on an elliptic curve. No such algorithm is known in general, although there are some special cases where we have a simple deterministic algorithm (see Exercise 1 of §2). The square root algorithm above is not deterministic polynomial-time unless one assumes the Riemann Hypothesis, because of the difficulty in deterministically finding a non-square u. If q is a prime p that is not congruent to 1 modulo 16, then [Schoof 1985] contains a deterministic polynomial-time square root algorithm that does not assume the Riemann Hypothesis.

But the main obstacle to a deterministic polynomial-time algorithm for finding a point on E is not the problem of taking the square root of $f(x)$. Rather, it is finding $x \in \mathbb{F}_q$ such that $f(x)$ is a square. Although about 50% of the elements x have this property, no efficient deterministic way is known to find such an x except in some special cases. All deterministic methods require exponential time. It is remarkable that no one knows how to find a point on an elliptic curve (other than the point at infinity) in subexponential time.

On the other hand, if we are satisfied with probabilistic methods (as we always are if we work in practical cryptography), then we have nothing to worry about.

Exercises for §1

1. Show that a linear change of variables can be used to transform the left side of equation (1) to the form
(a) Y^2 if the characteristic of the field $\mathbb{F}$ is not 2; and
(b) $Y^2 + XY$ if char($\mathbb{F}$) = 2 and the XY-term in equation (1) is nonzero.

2. If char($\mathbb{F}$) = 2, show that there is no elliptic curve (1) with $a_1 = a_3 = 0$.

3. (a) In the case when char($\mathbb{F}$) $\neq$ 2, suppose that equation (1) has been transformed as in Exercise 1(a) so that $a_1 = a_3 = 0$. Show that the equation defines an elliptic curve (in other words, a smooth curve) if and only if the cubic polynomial on the right has no multiple roots.
(b) In the case when char($\mathbb{F}$) = 2 and either a_1 or a_3 (but not both) is nonzero, give simple conditions for equation (1) to define an elliptic curve. Also show that if $a_1 \neq 0$, then without loss of generality we may suppose that $a_4 = 0$; while if $a_3 \neq 0$, then we may suppose that $a_2 = 0$.

4. Show how the proposition in §1.2 can be used to prove the associative law for $E(\mathbb{R})$.

5. How many points P such that $nP = O$ are there on an elliptic curve over (a) $\mathbb{C}$? (b) $\mathbb{R}$?

6. Let P be a point on an elliptic curve of the form (3) over $\mathbb{R}$. Give a geometric condition that is equivalent to P being a point of order (a) 2; (b) 3; (c) 4.

7. On the elliptic curve $Y^2 = X^3 - 36X$ let $P = (-3, 9)$ and $Q = (-2, 8)$. Find $P + Q$ and $2P$.

8. Each of the following points has finite order on the given elliptic curve over $\mathbb{Q}$. In each case, find the order of P.

(a) $P = (0, 16)$ on $Y^2 = X^3 + 256$.

(b) $P = (\frac{1}{2}, \frac{1}{2})$ on $Y^2 = X^3 + \frac{1}{4}X$.

(c) $P = (3, 8)$ on $Y^2 = X^3 - 43X + 166$.

9. Let E be a curve (3) defined over the rational numbers. For simplicity, suppose that $a, b \in \mathbb{Z}$. Let P be a point on $E(\mathbb{Q})$. Find a bound in terms of k for the logarithm of the denominator of the x-coordinate of $2^k P$.

10. Let E be either (a) the curve $Y^2 = X^3 - X$ defined over the field $\mathbb{F}_q$, where $q \equiv 3 \pmod 4$, or else (b) the curve $Y^2 + Y = X^3$ defined over the field $\mathbb{F}_q$, where $q \equiv 2 \pmod 3$. In both cases show that one has an elliptic curve (that is, the curve is smooth); prove that $N_1 = q + 1$; and find formulas for N_r.

11. Let $E/\mathbb{F}_2$ be the elliptic curve $Y^2 + Y = X^3$, and let $q = 2^r$.

(a) Express the coordinates of $-P$ and $2P$ in terms of the coordinates of P.

(b) Show that every $P \in E(\mathbb{F}_{16})$ (except for O) has order 3.

(c) Show that every $P \in E(\mathbb{F}_{16})$ is actually in $E(\mathbb{F}_4)$. Then use Hasse's theorem with $q = 4$ and with $q = 16$ to determine the number of $\mathbb{F}_{16}$-points. Your answer should agree with the formula for N_4 in (17).

12. Let $E/\mathbb{F}_p$ have equation $Y^2 + Y = X^3 - X + 1$, where $p = 2$ or 3. Show that $N_1 = 1$, and find a simple formula for N_r.

13. Find an elliptic curve E defined over $\mathbb{F}_4$ that has only one $\mathbb{F}_4$-point (the point at infinity). Find a simple formula for N_r in that case. Show that one has $(2^r - 1)P = O$ for all $P \in E(\mathbb{F}_{4^r})$.

14. Given an l-bit integer n and a point $P \in E(\mathbb{F}_q)$, where q is a k-bit prime power, prove that nP can be computed in time $O(k^2 l)$.

15. In the notation of Corollary 1.1, find a recursive relation expressing N_{r+1} in terms of N_r and N_{r-1} that can be used to compute the sequence N_r extremely rapidly once you know a.

16. For $a = 0$ or 1, let E_a be the elliptic curve $Y^2 + XY = X^3 + aX^2 + 1$ over $\mathbb{F}_2$. Find $\#E_a(\mathbb{F}_2)$ and $Z(E_a/\mathbb{F}_2; T)$ for $a = 0, 1$. Using Exercise 15, show that $\#E_a(\mathbb{F}_{2^r})$ is four times a prime when $a = 0$, $r = 5, 7, 13$, and is twice a prime when $a = 1$, $r = 3, 5, 7, 11$. It is easy by computer to find larger prime values of r for which $\#E_a(\mathbb{F}_{2^r})/\#E_a(\mathbb{F}_2)$ is a prime (see also §3.2 below). Such curves are suitable for elliptic curve cryptography, in part because they lend themselves to especially efficient computation of multiples of points (see [Solinas 1997]).

17. Prove that if $\mathbb{F}$ is a finite field of characteristic 2, then $\#E(\mathbb{F})$ is odd in the supersingular case (13) and even in the nonsupersingular case (12). Conclude that the coefficient a in the numerator of $Z(E/\mathbb{F}; T)$ is even in the supersingular case and odd in the nonsupersingular case.

§ 2. Elliptic Curve Cryptosystems

2.1 History

Elliptic curve cryptosystems were proposed in 1985 independently by Victor Miller and by me. The two advantages that we saw were (1) the greater flexibility in choosing the group (that is, for each prime power q there is only one multiplicative group $\mathbb{F}_q^*$, but there are many elliptic curve groups $E/\mathbb{F}_q$), and especially (2) the absence of subexponential time algorithms to break the system if E is suitably chosen.

At first, elliptic curve cryptography seemed like the sort of notion that would be of practical utility only in the distant future, if at all. However, as often happens in cryptography, the distant future came quickly. Now, a little more than a decade later, many people have developed usable implementations. Some of the most advanced work in this field (both practical and theoretical) has been done in Waterloo, Canada by an interdisciplinary group headed by S. A. Vanstone.

A few years after the invention of elliptic curve cryptosystems, Menezes, Okamoto, and Vanstone [1993] found a new way to tackle the discrete log problem (see Definition 2.1 below) upon which the security of elliptic curve cryptosystems is based. Namely, given an elliptic curve E defined over $\mathbb{F}_q$, they used the Weil pairing (see §III.8 of [Silverman 1986]) to imbed E into the multiplicative group of some extension $\mathbb{F}_{q^k}$. This reduces the problem to the discrete log problem in $\mathbb{F}_{q^k}^*$ (see §4 of Chapter 1). However, in order for this to be of any use, the extension degree k must be small. Essentially the only elliptic curves for which k is small are the supersingular ones. These include a few simple equations, such as the examples in Exercise 10 of §1, and also all equations of the form (13) in characteristic 2; however, the vast majority of elliptic curves are nonsupersingular. For them, the Menezes–Okamoto–Vanstone reduction almost never leads to a subexponential time algorithm (see [Balasubramanian and Koblitz 1998]).

Thus, the basic open question in elliptic curve cryptography is whether or not one can find a subexponential time algorithm for the discrete log problem on some class of nonsupersingular elliptic curves. At present no one seems to have any idea how to do this.

Meanwhile, because of progress in computing finite field discrete logarithms and in factoring integers, the key sizes necessary in order for the most popular public key systems to be secure are growing substantially. Thus, Odlyzko's article [Odlyzko 1995] on this subject concluded with the following sentence:

> "It might therefore be prudent to consider even more seriously elliptic curve cryptosystems."

2.2 Key Exchange and Message Transmission

One of the most attractive uses of a public key cryptosystem is for key exchange (where actual message transmission will be done by an unrelated private key system). The key can be any more-or-less "random" integer that the two users Alice and Bob agree upon but no one else knows. The unique feature of public key cryptography for key exchange is that Alice and Bob can arrive at their common key using only public, unencrypted communication.

The first public key cryptosystem was the Diffie–Hellman key exchange [Diffie and Hellman 1976] (see §4 of Chapter 1). It can be adapted for elliptic curves as follows. First note that a "random" point on an elliptic curve E can serve as a key, since Alice and Bob can agree in advance on a method to convert it to an integer (for example, they can take the image of its x-coordinate under some agreed upon simple map from $\mathbb{F}_q$ to the natural numbers).

So suppose that E is an elliptic curve over $\mathbb{F}_q$, and Q is an agreed upon (and publicly known) point on the curve. Alice secretly chooses a random integer k_A and computes the point $k_A Q$, which she sends to Bob. Likewise, Bob secretly chooses a random k_B, computes $k_B Q$, and sends it to Alice. The common key is $P = k_A k_B Q$. Alice computes P by multiplying the point she received from Bob by her secret k_A; Bob computes P by multiplying the point he received from Alice by his secret k_B. An eavesdropper who wanted to spy on Alice and Bob would have to determine $P = k_A k_B Q$ knowing Q, $k_A Q$, and $k_B Q$, but not k_A or k_B. The eavesdropper's task is called the "Diffie–Hellman problem for elliptic curves".

It is not hard to modify the Diffie–Hellman protocol for the purpose of message transmission, using an idea of ElGamal [1985a]. Suppose that the set of message units has been imbedded in E in some agreed upon way (see Exercises 2–4 below), and Bob wants to send Alice a message $M \in E$. Alice and Bob have already exchanged $k_A Q$ and $k_B Q$ as in Diffie–Hellman. Bob now chooses another secret random integer l, and sends Alice the pair of points $(lQ, M + l(k_A Q))$. To decipher the message, Alice multiplies the first point in the pair by her secret k_A and then subtracts the result from the second point in the pair.

The Diffie–Hellman and ElGamal systems can be broken if one can solve the "discrete log problem" in the group E.

Definition 2.1. The *discrete logarithm problem in the group G to the base* $g \in G$ is the problem, given $y \in G$, of finding an integer x such that $g^x = y$ ($xg = y$ when the group operation in G is written additively), provided that such an integer exists (in other words, provided that y is in the subgroup generated by g). Thus, in the case $G = E$, the *elliptic curve discrete logarithm problem to the base* $Q \in E$ is the problem, given $P \in E$, of finding an integer x such that $P = xQ$ if such x exists.

It is easy to see that the Diffie–Hellman problem can be solved if the discrete log problem can be. Namely, the eavesdropper, who knows Q and $k_A Q$, finds the secret k_A and then has broken the cipher. The converse – the assertion that the

Diffie–Hellman problem is *equivalent* to the discrete log problem – is a conjecture but has not been proved. For the latest partial results supporting the conjectural equivalence of these two problems, see [Boneh and Lipton 1996] and [Maurer and Wolf 1998].

The system that Diffie and Hellman originally proposed used the multiplicative group $G = \mathbb{F}_q^*$. The discrete logarithm problem in the multiplicative group of a well chosen finite field is hard: in practice, it seems to require about the same amount of time as factorization of an integer of approximately the same size as the finite field (see [van Oorschot 1992]). However, as in the case of factorization, there are many subexponential time algorithms to do this. More precisely, most good discrete log algorithms have running time of the form $L_q(1/2) = \exp\big(O(\sqrt{\ln q \ln \ln q})\big)$; and with the discovery and development of the "number field sieve" (see [Lenstra and Lenstra 1993] and [Gordon 1993 and 1995]), the heuristic asymptotic running time (at least in the case when $q = p$ is a prime) has been reduced to $L_q(1/3) = \exp\big(O(\sqrt[3]{\ln q \ln \ln^2 q})\big)$.

However, no subexponential time algorithm is known in the case of elliptic curves (except for supersingular ones, see §2.1). The only methods available for finding discrete logs on $E/\mathbb{F}_q$ are the methods that apply to arbitrary groups. All of them have running time of the form $L_q(1) = q^{O(1)} = \exp\big(O(\ln q)\big)$, provided that $\#E$ is divisible by a large prime ("large" means that its order of magnitude is not much less than that of q).

2.3 Discrete Log Algorithm in Groups of Smooth Order

Definition 2.2. Let B be a positive real number. An integer is said to be B-*smooth* if it is not divisible by any prime greater than B.

If the order of our group G is B-smooth for a reasonably small B, then discrete logarithms in G can be efficiently computed by the following method of Silver–Pohlig–Hellman [Hellman and Pohlig 1978].

Let G be a group of order $\#G = \prod p_i^{s_i}$. We shall write the group law additively, and let O denote the identity element. Suppose that $\#G$ is B-smooth; in other words $p_i \leq B$ for all i. If the bound B is fairly small, then one can use the following algorithm to find the discrete logarithm of $y \in G$ to the base g. First we find the exact order of g. This can be done by computing $(\#G/p_i)g$ for the different p_i, and then $(\#G/p_i^2)g$ whenever $(\#G/p_i)g = O$, and so on, until we find the smallest $N = \prod p_i^{r_i}$ such that $Ng = O$.

Our task is to find a positive integer $x < N$ such that $xg = y$. If no such x exists, then the algorithm that follows will break down, and we will know that there is no solution.

Our method is to find x modulo p^r, where p^r is one of the prime powers in the factorization of N, and then use the Chinese Remainder Theorem (see Exercise 9 in §3 of Chapter 2) to find x modulo $N = \prod p^r$. So suppose that p is a fixed prime divisor of N, and let $x \equiv x_0 + x_1 p + \cdots + x_{r-1} p^{r-1} \pmod{p^r}$, where $0 \leq x_i \leq p-1$ for $i = 0, 1, \ldots, r-1$.

To find the unknown digit x_0, we multiply both sides of the equality $xg = y$ by $N' = N/p$, obtaining $x_0(N'g) = N'y$. We could simply try the p different possibilities for x_0 until we find the one for which the last equality holds; if no such $x_0 \in \{0, 1, \ldots, p - 1\}$ exists, that means that y is not in the subgroup generated by g. This procedure requires $O(p)$ steps. If p is fairly large, we might instead want to use Shanks' "baby–step–giant–step" method, which requires only $O(\sqrt{p})$ steps. Namely, we let $b = [\sqrt{p}] + 1$, and we write our unknown x_0 in the form $x_0 = ib - j$, where $1 \le i, j \le b$. We compute the b values $i(bN'g)$, $1 \le i \le b$, and the b values $(N'y) + j(N'g)$, $1 \le j \le b$, and we compare the two lists. When we have the match $ibN'g = N'y + jN'g$, we know that $x_0 = ib - j$.

Once we know x_0, we find x_1 by setting $N'' = N/p^2$ and considering the equality $(x_0 + x_1 p)(N''g) = N''y$, i.e., $x_1(N'g) = N''(y - x_0 g)$. We use the same procedure that we used to find x_0. We continue this process, inductively finding $x_2, \ldots, x_{r-1}$. Once we know x modulo p^r for all $p|N$, we use the Chinese Remainder Theorem to find x, $1 \le x < N$. This completes the description of the algorithm.

Remarks. 1. Unlike index-calculus algorithms for discrete logs in $\mathbb{F}_q^*$, in which most of the work is in precomputations (which are not repeated if one wants another discrete log in the same field), the above algorithm must be substantially repeated for each individual discrete logarithm.

2. Because of the importance of the discrete logarithm problem on an elliptic curve, much effort has been devoted to obtaining even minor increases in speed. A method based on Pollard's ρ-method [Pollard 1978], which has been efficiently parallelized in [Van Oorschot and Wiener 1994, 1998], is somewhat faster than the above combination of Silver–Pohlig–Hellman and baby–step–giant–step.

3. Whenever the security of a cryptosystem is based on the discrete logarithm problem in a group G, to thwart attacks by any of these algorithms it is important to choose G so that it has non-smooth order – in other words, so that $\#G$ is divisible by a very large prime. In practice, all known algorithms that work in an arbitrary group become infeasible if $\#G$ is divisible by a prime of 40 or more decimal digits.

2.4 Digital Signature

We now describe the elliptic curve analogue (ECDSA) of the U.S. government Digital Signature Algorithm (see §4 of Chapter 1). The ECDSA is currently being studied by the standards committees of several professional organizations, and it may soon be adopted as a digital signature standard that can be used as an alternative to the DSA.

ECDSA Key Generation. For simplicity, we shall use elliptic curves defined over a prime field $\mathbb{F}_p$, although the construction can easily be adapted to other finite fields as well. Let E be an elliptic curve defined over $\mathbb{F}_p$, and let P be a point of prime order q in $E(\mathbb{F}_p)$; these are system-wide parameters. (Note that here, as in the DSA in §4 of Chapter 1, q denotes *not* a power of p, but rather a different prime number. Unlike in the DSA, where q is much smaller than p, in the ECDSA q is about the same size as p.) Each user Alice selects a random integer x in the interval $1 < x < q - 1$ and computes $Q = xP$. Alice's public key is Q; her private key is x.

ECDSA Signature Generation. To sign a message m, Alice does the following:

1) She selects a random integer k in the interval $1 < k < q - 1$.
2) She computes $kP = (x_1, y_1)$ and $r = x_1 \bmod q$ (that is, x_1 is regarded as an integer between 0 and $p - 1$, and r is taken to be its least non-negative residue modulo q). If $r = 0$, she returns to step 1). (If $r = 0$, then the signing equation $s = k^{-1}(H(m) + xr) \bmod q$ does not involve the private key x; hence, 0 is not a suitable value for r.)
3) She computes $k^{-1} \bmod q$.
4) She computes $s = k^{-1}(H(m) + xr) \bmod q$, where $H(m)$ is the hash value of the message. If $s = 0$, she returns to step 1). (If $s = 0$, then $s^{-1} \bmod q$, which is required in step 3) of signature verification, does not exist. Note that if k is chosen at random, then the probability that either $r = 0$ or $s = 0$ is negligibly small.)
5) The signature for the message m is the pair of integers (r, s).

ECDSA Signature Verification. To verify Alice's signature (r, s) of the message m, Bob should do the following:

1) Obtain an authenticated copy of Alice's public key Q.
2) Verify that r and s are integers in the interval $[1, q - 1]$.
3) Compute $w = s^{-1} \bmod q$ and $H(m)$.
4) Compute $u_1 = H(m)w \bmod q$ and $u_2 = rw \bmod q$.
5) Compute $u_1 P + u_2 Q = (x_0, y_0)$ and $v = x_0 \bmod q$.
6) Accept the signature if and only if $v = r$.

The basic difference between ECDSA and DSA is in the generation of r. The DSA does this by taking the random power $(\alpha^k \bmod p)$ and reducing it modulo q, thus obtaining an integer in the interval $[1, q - 1]$. (Recall that in DSA q is a 160-bit prime divisor of $p - 1$, and α is an element of order q in $\mathbb{F}_p^*$.) The ECDSA generates the integer r in the interval $[1, q - 1]$ by taking the x-coordinate of the random multiple kP and reducing it modulo q.

To obtain a security level similar to that of DSA, the parameter q should have about 160 bits. If this is the case, then DSA and ECDSA signatures have the same bitlength (320 bits).

Instead of using E and P as system-wide parameters, we could fix only the underlying finite field $\mathbb{F}_p$ for all users, and let each user select her own elliptic curve E and point $P \in E(\mathbb{F}_p)$. In this case, the defining equation for E, the coordinates of the point P, and the order q of P must also be included in the user's public key. If the underlying field $\mathbb{F}_p$ is fixed, then hardware and software can be built to optimize computations in that field. At the same time, there are an enormous number of choices of elliptic curve E over the fixed $\mathbb{F}_p$.

Exercises for § 2

1. In the ECDSA, explain why (a) Bob expects the x-coordinate of $u_1 P + u_2 Q$ to agree modulo q with r, and (b) if they do agree, then he should be satisfied that it was really Alice who sent the message.

2. If $E/\mathbb{F}_q$ is either of the types of elliptic curves in Exercise 10 of §1, describe a deterministic method of imbedding plaintext as points on the curve. If $E/\mathbb{F}_q$ is some other type of elliptic curve, where q is odd, describe a probabilistic method of doing this.

3. Let $q = 2^r$, and let $\{\beta_1, \ldots, \beta_r\}$ be a basis of $\mathbb{F}_q$ over $\mathbb{F}_2$. Let $\mathrm{Tr}(z) = \sum_{i=0}^{r-1} z^{2^i}$ be the trace function. By Exercise 13 in §2 of Chapter 3, $\mathrm{Tr}(z)$ is an additive map from $\mathbb{F}_q$ to $\mathbb{F}_2$ that takes the value 1 for $q/2$ elements of $\mathbb{F}_q$ and the value 0 for the other $q/2$ elements of $\mathbb{F}_q$.
(a) For $z \in \mathbb{F}_q$ show that the equation $u^2 + u = z$ can be solved for $u \in \mathbb{F}_q$ if and only if $\mathrm{Tr}(z) = 0$. If $\mathrm{Tr}(z) = 0$, describe an algorithm for finding u.
(b) If $\{\beta_1, \ldots, \beta_r\}$ is a *normal* basis (that is, $\beta_i = \beta^{2^{i-1}}$, $i = 1, \ldots, r$, for some fixed $\beta \in \mathbb{F}_q$), give a simple criterion for z to have trace zero, and describe a very easy way to solve the equation $u^2 + u = z$ in that case.
(c) Let E be an elliptic curve of the form (12) or (13) defined over $\mathbb{F}_q$. Using part (a), describe a probabilistic algorithm for finding points on E.

4. Consider an elliptic curve E with equation (12) over a small field $\mathbb{F}_q$, $q = 2^r$, where you are free to vary the $a_2, a_6 \in \mathbb{F}_q$. Let f denote a prime number (which you are also free to vary). Let $N = \#E(\mathbb{F}_q)$, and let $N_f = \#E(\mathbb{F}_{q^f})$. Describe an algorithm for finding $a_2, a_6 \in \mathbb{F}_q$ and a prime f such that N_f/N is a prime number, if such a_2, a_6, and f exist.

5. Suppose that the best algorithm for discrete logarithms in $\mathbb{F}_q^*$ requires time $\exp\left(O\left((\ln q)^{1/3}(\ln \ln q)^{2/3}\right)\right)$. Show that the Menezes–Okamoto–Vanstone reduction does not give a subexponential time algorithm if $k \geq (\ln q)^2$, where k is the degree of the extension field of $\mathbb{F}_q$ in which the elliptic curve group is imbedded. Thus, if l is a prime dividing $\#E(\mathbb{F}_q)$ that has the same order of magnitude as q, and if $l \nmid q^k - 1$ for $k < \ln^2 q$, then the Menezes–Okamoto–Vanstone reduction does not lead to a subexponential time algorithm for the discrete logarithm in the group of order l in $E(\mathbb{F}_q)$.

6. Show how Cathy could mount an adaptive chosen-ciphertex attack on the El-Gamal encryption system in §2.2 (see Exercise 11 of §3 of Chapter 5).

§ 3. Elliptic Curve Analogues
of Classical Number Theory Problems

There are basically three approaches to choosing an elliptic curve for a crypto-system. In each case one looks for a curve whose order $\#E$ has a very large prime factor, and in each case the question of the likelihood of encountering such a curve leads to some interesting conjectures that are supported by heuristic ar-guments and computational evidence. However, proving them remains a difficult unsolved problem.

3.1 Fix a "Global" Elliptic Curve and Vary the Prime

For example, let E be an elliptic curve $Y^2 = f(X) = X^3 + aX + b$ defined over the field $\mathbb{Q}$ of rational numbers. If p is any odd prime not dividing the denominators of the coefficients or the discriminant of $f(X)$, then one can consider the elliptic curve E over $\mathbb{F}_p$ that is obtained by simply reducing the coefficients modulo p. That elliptic curve will always contain as a subgroup the image of the torsion subgroup E_{tors} of the curve over $\mathbb{Q}$ (see §1.5). But one expects that in many cases the quotient will have prime order.

QUESTION. For a fixed curve E over $\mathbb{Q}$, what can be said about the probability as p varies that

$$\frac{\#E \bmod p}{\#E_{\text{tors}}}$$

is a prime number? Can one prove (for any fixed E) that there are infinitely many p for which this number is prime?

This question is analogous to a classical unsolved problem of number theory. Namely, if instead of E we take the multiplicative semigroup of nonzero integers, which has torsion subgroup $\{\pm 1\}$, then an analogous question is: As p varies, what can be said about the probability that

$$p_1 = \frac{p-1}{2} = \frac{\#\mathbb{F}_p^*}{\#\{\pm 1\}}$$

is prime? Are there infinitely many such "Sophie Germain primes" p_1 for which $p = 2p_1 + 1$ is prime?*
The question about Sophie Germain primes is of interest when using a Diffie–Hellman type cryptosystem in the multiplicative group of a prime field $\mathbb{F}_p$, and the analogous elliptic curve question given above is of interest when using an elliptic curve cryptosystem. In both cases one needs the order of the group to be divisible

---

* In 1823 Sophie Germain proved the so-called "first case" of Fermat's Last Theorem for prime exponents p_1 for which $2p_1 + 1$ is prime. This was the first major result on Fermat's Last Theorem for a large class of exponents.

by a large prime so that the Silver–Pohlig–Hellman method (see §2.3) cannot be used to find discrete logs.

It should be noted that the denominator $\#E_{\text{tors}}$ in the elliptic curve question is often 1, and in any case it cannot be much larger than in the Sophie Germain prime question. According to a deep result of Mazur, there are at most 16 torsion points on an elliptic curve over $\mathbb{Q}$ (see [Mazur 1977]).

For a discussion of conjectures that would answer the above question for elliptic curves, see [Koblitz 1988].

Another natural question in this context is the following. Suppose that P is a point of infinite order in the group of the elliptic curve E over $\mathbb{Q}$. As p varies, what is the probability that the image modulo p of P generates E modulo p? For results on this problem, see [Gupta and Murty 1986]. This question is the analogue of a classical problem of E. Artin, who conjectured formulas for the probability that a fixed integer a (such as 2 or 10) generates $\mathbb{F}_p^*$ as p varies. Note that a generates $\mathbb{F}_p^*$ if and only if the base-a expansion of $1/p$ has the longest possible period $p - 1$.

3.2 Fix an Elliptic Curve over a Small Field $\mathbb{F}_q$ and Then Consider It over $\mathbb{F}_{q^r}$ As r Varies

Since $E(\mathbb{F}_{q^{r'}})$ is a subgroup of $E(\mathbb{F}_{q^r})$ whenever $r'|r$, large prime factors of $\#E(\mathbb{F}_{q^r})$ are more likely to occur when r is prime than when r is composite. In the case of prime r, the best one can hope for is that

$$\frac{\#(E(\mathbb{F}_{q^r}))}{\#(E(\mathbb{F}_q))} = \left| \frac{\alpha^r - 1}{\alpha - 1} \right|^2$$

is prime. (Here α is a reciprocal root of the numerator of $Z(E/\mathbb{F}_q; T)$.)

QUESTION. For fixed $E/\mathbb{F}_q$, what is the probability as r varies that the above number is prime? Can one ever prove that there are infinitely many r such that it is prime?

Virtually nothing is known about this question. It is analogous to the classical Mersenne prime problem, as we see by replacing α by 2.

3.3 Fix the Field of Definition $\mathbb{F}_q$ and Vary the Coefficients

According to Hasse's theorem (Corollary 1.2), $\#E$ falls in a rather small interval around $q + 1$, namely,

$$[q + 1 - 2\sqrt{q}, q + 1 + 2\sqrt{q}] .$$

As E ranges over all elliptic curves defined over $\mathbb{F}_q$, the number $\#E$ is distributed fairly uniformly in this interval, except that the density drops off near the endpoints (see [Waterhouse 1969] and [Lenstra 1987]). Thus, the probability that $\#E$ is prime (or has a prime factor greater than some lower bound) is essentially the same as the probability that a random integer in an interval of the form $[q, q + c\sqrt{q}]$ (c a constant) has this property. But unfortunately, at present almost nothing can be proved about the occurrence of primes in such "short" intervals. It is not even known whether there exists a c such that the interval $[q, q + c\sqrt{q}]$ always contains at least one prime as $q \longrightarrow \infty$.

Exercises for § 3

1. State a number theory problem similar to the Sophie Germain prime problem that relates to the cryptographic suitability of curves of the type in Exercise 10 of §1.

2. Let E be the elliptic curve $Y^2 + Y = X^3 - X$ defined over $\mathbb{Q}$, and let $P = (0, 0)$. It can be shown that $E(\mathbb{Q})$ is an infinite cyclic group generated by P. Find an example of a prime p such that the curve $E(\mathbb{F}_p)$ given by the same equation considered over $\mathbb{F}_p$ is *not* generated by the point $(0, 0)$.

3. Let E be the curve $Y^2 + Y = X^3 - X + 1$ over $\mathbb{F}_2$ (see Exercise 12 of §1). State a problem relating to cryptographic suitability of $E(\mathbb{F}_{2^r})$ that is closely analogous to the Mersenne prime question. (However, regardless of the answer to this question, the Menezes–Okamoto–Vanstone reduction leads to a subexponential algorithm, because the curve is supersingular. The same remark applies to Exercise 1.)

§ 4. Cultural Background: Conjectures on Elliptic Curves and Surprising Relations with Other Problems

4.1 Congruent Numbers

The following "congruent number problem" has been around since ancient times (see Chapter XVI of [Dickson 1952] and Section D27 of [Guy 1981]): Given a natural number N, does there exist a right triangle with rational sides whose area is N? Is there an easy way to determine whether an arbitrary N is such a *congruent number*? Because of the famous 3–4–5 triangle, any high school student can see that $N = 6$ is a congruent number. So is $N = 5$, although not every high school student would be able to show that: the simplest example of a right triangle with rational sides and area 5 is the $1\frac{1}{2}$–$6\frac{2}{3}$–$6\frac{5}{6}$ triangle. It turns out that 1, 2, 3, and 4 are not congruent numbers.

It is not hard to show that N is a congruent number if and only if the elliptic curve $Y^2 = X^3 - N^2 X = X(X - N)(X + N)$ has a nontrivial point, where "nontrivial" means excluding the point at infinity and the other three points of order two: $(0, 0)$ and $(\pm N, 0)$. For instance, in the case $N = 6$ (see Exercise 7 of §1) the point $P = (-3, 9)$, which is a point of infinite order on the curve $Y^2 = X^3 - 36X$, corresponds to the 3–4–5 right triangle.

For more information on the congruent number problem see [Koblitz 1993] and [Tunnell 1983].

4.2 Fermat's Last Theorem

In 1985 Gerhard Frey suggested that if $A^p + B^p = C^p$ were a counterexample to Fermat's Last Theorem, then the elliptic curve

$$Y^2 = X(X - A^p)(X + B^p)$$

would have a very surprising property. Its discriminant would be

$$-\left(A^p B^p (A^p + B^p)\right)^2 = -(ABC)^{2p} \; ,$$

so every prime factor in this discriminant would occur to a very large power. Frey thought that it would then have to violate the so-called *Taniyama conjecture*. K. Ribet was able to prove that Frey's hunch was correct [Ribet 1990]; then, working intensively for many years, A. Wiles (partly in joint work with R. Taylor) proved that no such curve can violate the Taniyama conjecture, and hence there can be no counterexample to Fermat's Last Theorem.

A more detailed discussion of this dramatic story would take us too far afield. See [Faltings 1995] for a concise summary of Wiles' proof.

4.3 The Birch–Swinnerton-Dyer Conjecture

Whenever we have an elliptic curve E as in (3) defined over the rational numbers (that is, $a, b \in \mathbb{Q}$), we can consider it modulo p for any prime p that does not divide either the denominator of a or b or the discriminant $-(4a^3 + 27b^2)$. This is a curve defined over $\mathbb{F}_p$; as in §3.1, we shall denote it "E mod p". For a fixed E over $\mathbb{Q}$ and variable p, let N_p denote #(E mod p). (This use of the subscript with N is different from that in Corollary 1.1.)

Recall from §1 that $N_p = (\overline{\alpha}_p - 1)(\alpha_p - 1)$, where α_p and $\overline{\alpha}_p$ are the quadratic imaginary numbers of absolute value $\sqrt{p}$ that one gets from factoring the numerator of (15).

As p increases, suppose that we want to get an idea of whether or not N_p tends to be toward the right end of the interval $[p + 1 - 2\sqrt{p}, p + 1 + 2\sqrt{p}]$ (see (18)), that is, whether or not there tend to be more points on the curve than one would expect if the right side of equation (3) (modulo p) had exactly a 50% chance of producing a quadratic residue as p and x vary. We might expect that if our original curve over $\mathbb{Q}$ has infinitely many points – that is, if its rank r is positive (see §1.5) – then these points would be a plentiful source of mod-p points, and N_p would tend to be large; whereas if $r = 0$, then N_p would straddle both sides of $p + 1$ equally. This is the intuitive idea of the (weak) *Birch–Swinnerton-Dyer conjecture* (see [Birch and Swinnerton-Dyer 1963, 1965] and [Cassels 1966]).

To measure the relative size of N_p and p as p varies, let us form the product $\prod_p \frac{p}{N_p}$. Because

$$N_p = (\overline{\alpha}_p - 1)(\alpha_p - 1) = (\frac{p}{\alpha_p} - 1)(\frac{p}{\overline{\alpha}_p} - 1) = \frac{1}{p}(p - \alpha_p)(p - \overline{\alpha}_p) \,,$$

it follows that our product over p of the ratios p/N_p can be written as follows:

$$\prod_p \frac{p}{N_p} = \prod_p \frac{p^2}{(p - \alpha_p)(p - \overline{\alpha}_p)} = \prod_p \frac{1}{\left(1 - \frac{\alpha_p}{p}\right)\left(1 - \frac{\overline{\alpha}_p}{p}\right)} \,.$$

One might expect that this infinite product would converge to zero if N_p has a tendency to be significantly larger than p, and would converge to a nonzero value if N_p is equally likely to be above or below p. As it happens, this infinite product does not converge at all. However, it can be viewed as the value at $s = 1$ of the function

$$\prod_p \frac{1}{\left(1 - \frac{\alpha_p}{p^s}\right)\left(1 - \frac{\overline{\alpha}_p}{p^s}\right)} \,.$$

This infinite product can easily be shown to converge for any s with real part greater than $3/2$; and, like the Riemann zeta-function that it resembles, it can be analytically continued onto the rest of the complex plane. (The latter property is deep; it has been proved for a very broad class of elliptic curves over $\mathbb{Q}$, but remains a conjecture for elliptic curves not in this class.) The Birch–Swinnerton-Dyer conjecture states that this function vanishes at $s = 1$ if and only if the rank r of the group of E over $\mathbb{Q}$ is greater than zero, and that, moreover, its order of vanishing at $s = 1$ is equal to r. The conjecture further says that the leading coefficient in the Taylor expansion at $s = 1$ can be expressed in terms of certain number-theoretic invariants of E. In the 1980's important partial results were proved in support of this remarkable conjecture, but in its most general form it remains a very difficult open problem.

4.4 The Sato–Tate Distribution

There is another question that naturally arises when we fix an elliptic curve E over $\mathbb{Q}$, let p vary, and study

$$N_p = \#(E \bmod p) = p + 1 - \alpha_p - \overline{\alpha}_p = p + 1 - 2\sqrt{p}\,\mathrm{Re}\left(\frac{\alpha_p}{\sqrt{p}}\right) \,.$$

If we choose α_p to have non-negative imaginary part, then $p^{-1/2}\alpha_p$ is on the upper unit semicircle. Let $\theta_p \in [0, \pi]$ be its argument. According to a conjecture of Sato and Tate (see [Tate 1965]), if E does not have complex multiplication (see the end of §1.4), then as p increases the θ_p are distributed like the function $\frac{2}{\pi}\sin^2\theta$. Equivalently, the probability that $p^{-1/2}\alpha_p$ has argument between θ and $\theta + \Delta\theta$ is proportional (in the limit for large p and small $\Delta\theta$) to the area under the segment between θ and $\theta + \Delta\theta$ of the graph of the semicircle function $y = \sqrt{1 - x^2}$. (See the drawing on the next page, where the shaded area is $\approx y\Delta x \approx \sin\theta(\frac{d}{d\theta}\cos\theta)\Delta\theta = \sin^2\theta|\Delta\theta|$, since $y = \sin\theta$ and $x = \cos\theta$.)

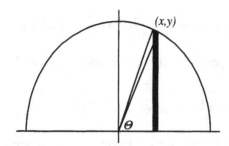

As far as I know, neither the Birch–Swinnerton-Dyer conjecture nor the Sato–Tate conjecture has had any application to cryptography. However, the somewhat different question of how $\#E(\mathbb{F}_p)$ is distributed in the interval $[p + 1 - 2\sqrt{p}, p + 1 + 2\sqrt{p}]$ for fixed p and variable coefficients $a, b \in \mathbb{F}_p$ is important in elliptic curve factorization [Lenstra 1987], elliptic curve primality testing [Goldwasser and Kilian 1986], and the design of elliptic curve cryptosystems (see §3.3).

4.5 Back to Cryptography:
Why No One Knows How to Find a "Factor Base"

Example 4.1. Let E be the elliptic curve $Y^2 + Y = X^3 - X$ over the field $\mathbb{Q}$ of rational numbers (see Exercise 2 of §3). Its group of rational points is infinite cyclic and is generated by $P = (0, 0)$. Using the formulas in §1, it is easy to see that $2P = (1, 0)$ and that, if the coordinates of nP are (x, y) for $n \geq 2$, then the coordinates of $(n+1)P$ are $\left(\left(\frac{y}{x}\right)^2 - x, -\left(\frac{y}{x}\right)^3 + y - 1\right)$. So we can quickly compute a table of the exact rational coordinates of nP for $1 \leq n \leq 50$. To get an idea of how fast the numerators and denominators of the coordinates grow, on the next page we tabulated the absolute value of the y-coordinate of nP for $7 \leq n \leq 50$.

Notice the parabolic appearance of this table. We see that the number of digits needed to express the coordinates of nP grows quadratically as a function of n. In other words, the *height* – defined as $\max\{|a|, |b|, |c|, |d|\}$ for $nP = (a/b, c/d)$ (with $x = a/b$ and $y = c/d$ written in lowest terms) – grows superexponentially: $\text{height}(nP) = e^{O(n^2)}$. It can be shown that this extremely rapid growth occurs for any $\mathbb{Q}$-point P of infinite order on an elliptic curve. In this connection see Exercise 9 of §1.

This means that there are very few points on E mod p that can be obtained by reduction modulo p of a $\mathbb{Q}$-point of E having small height. This is in striking contrast with the group $\mathbb{F}_p^*$, many of whose elements are obtained by reducing small integers modulo p. It is for this reason that most people are doubtful about the possibility of applying to the elliptic curve discrete log problem the index calculus methods that have been so successful in factoring integers and in finding discrete logs in $\mathbb{F}_q^*$. For more discussion of this question see [Miller 1986].

8
27
69
125
435
343
2065
64
3612
12167
28888
24389
43355
205379
2616119
2146689
28076979
30959144
332513754
274625
331948240
3574558889
8280062505
50202571769
641260644409
553185473329
18784454671297
4302115807744
318128427505160
578280195945297
10663732503571536
1469451780501769
663163334575107447
238670664494938073
8938035295591025771
13528653463047586625
58831063075349191045
487424450554237378792
316361137220162883230
343216282443844010111
435912379274109872312968
2035972062206737347698803
4197440172185492981774227
63061816101171948456692661
2181616293371330311419201915
4801616835579099275862827431
754388827236735824355996347601
495133617181351428873673516736
864938186463109226065267774743956
105831230775844387744754441796151
1220166832395032595688219182513256
80871745605559864852893980186125
253863219659861232674408424330433645
1417854835344178787714550500916300011
80758747641526362425597637684850206815
272173162387524688124210116071636697601
43588991327163432486545613110592404899171
34667409118422632032070193604742966179656
13092420486726785928582676734341299347115698
1911563313687376346904659714182809942964351
2189483446304543061371288657290348890980594400
4593462133720075853302175244236393876880522193
878446167724531925433595104275654803292506071599
214611697633277024760852115267283591567237967601
48072547288390781193922630252285326406309975292943
237844520166539043234240287233775865292113468549857
1241964389889828936323699134917565629213183077522469249
1583853196263084439374759692219941737511923840640000000
86882155212892274588612983069418029552410839095153302800
77199880835273348763199754318716659251711787046519688993
53875991960104970765283107373605344809310387645985049959200
112693242618274179956651080434126947419217320185504463599
4736963419597991005921784696415662858882109120520022240208657
37530201088428893504297348999259274784075840220078303129323057
5524905488265614668992892434873748844920757678725926478259255853
174093950939261851068856310479525754578138648900940338649752849
8988854703297735806987625854273224398026564032032776384182186736165
12789521106940379462098473081947860361167209079884962192477681132744
240664104114833357599838577779951121277275124199036386947622283276346814
122175911437214525606137783324557778424776691316115112703021731989593599
29214087094205221480230038284783574921809176572474206696449157526076526744
398745497990385157441648598714198500898417335456037004381292650442289963339
49253432232175037853939274934357858897623864482230655849646172771417862103019
1111599924813736177345961430101881701789905659558155903791530971080388326125
15592711215337616623740514518966145859285850662560621449737978910137952277603645
626031196063090566267061482108877864819511938185038642319827266816293763131719799
65343698144990446428357439135977881124804221113554492507243553294512904673973173265
159564798621271700005828929931002008441744804573070282618997694000714045237979692864

§5. Hyperelliptic Curves

In this section we shall give the main definitions and properties of hyperelliptic curves and their jacobians. Details and proofs can be found in the Appendix.

Let $\mathbb{F}$ be a finite field, and let $\overline{\mathbb{F}}$ denote its algebraic closure (see Definition 1.8 of Chapter 3).

5.1 Definitions

Definition 5.1. A *hyperelliptic curve C of genus g over $\mathbb{F}$ ($g \geq 1$)* is an equation of the form

$$C : v^2 + h(u)v = f(u) \qquad \text{in} \qquad \mathbb{F}[u, v] , \tag{20}$$

where $h(u) \in \mathbb{F}[u]$ is a polynomial of degree at most g and $f(u) \in \mathbb{F}[u]$ is a monic polynomial of degree $2g+1$. This curve must be smooth at all points $(x, y) \in \overline{\mathbb{F}} \times \overline{\mathbb{F}}$ that satisfy the equation $y^2 + h(x)y = f(x)$ (that is, no such points satisfy the partial derivative equations $2y + h(x) = 0$ and $h'(x)y - f'(x) = 0$).

Let $\mathbb{K}$ be a field containing $\mathbb{F}$. By a $\mathbb{K}$-point $P \in C$ we mean either the symbol ∞ (called the *point at infinity* on the curve C) or else a solution $(x, y) \in \mathbb{K} \times \mathbb{K}$ of the equation (20).

Definition 5.2. If $P = (x, y)$ is a $\mathbb{K}$-point of the hyperelliptic curve (20), we define its *opposite* $\widetilde{P}$ to be the other point with the same x-coordinate that satisfies the equation of the curve: $\widetilde{P} = (x, -y - h(x))$. If $P = \infty$, we take $\widetilde{P} = \infty$.

Definition 5.3. A *divisor* on C is a finite formal sum of $\overline{\mathbb{F}}$-points $D = \sum m_i P_i$. Its *degree* is the sum of the coefficients $\sum m_i$. If $\mathbb{K}$ is an algebraic extension of $\mathbb{F}$, we say that D is *defined over* $\mathbb{K}$ if for every automorphism σ of $\overline{\mathbb{F}}$ that fixes $\mathbb{K}$ one has $\sum m_i P_i^\sigma = D$, where P^σ denotes the point obtained by applying σ to the coordinates of P (and $\infty^\sigma = \infty$). Let $\mathbb{D}$ denote the additive group of divisors defined over $\mathbb{K}$ (where $\mathbb{K}$ is fixed), and let $\mathbb{D}^0$ denote the subgroup consisting of divisors of degree 0.

Definition 5.4. The *greatest common divisor* of $D = \sum m_i P_i \in \mathbb{D}^0$ and $D' = \sum m_i' P_i \in \mathbb{D}^0$ is defined to be $\left(\sum \min(m_i, m_i') P_i\right) - (*)\infty$, where the coefficient $(*)$ is chosen so that the greatest common divisor has degree 0.

Definition 5.5. Given a polynomial $G(u, v) \in \overline{\mathbb{F}}[u, v]$, we can consider $G(u, v)$ as a function on the curve (equivalently, as an element of the quotient ring $\overline{\mathbb{F}}[u, v]/(v^2 + h(u)v - f(u))$, see Example 4.1 in Chapter 3). This means that we lower the power of v in $G(u, v)$ by means of the equation of the curve until we have an expression of the form $G(u, v) = a(u) - b(u)v$. We let $(G(u, v)) = \left(\sum m_i P_i\right) - (*)\infty \in \mathbb{D}^0$ denote the *divisor of the polynomial function $G(u, v)$*, where the coefficient m_i is the "order of vanishing" of $G(u, v)$ at the point P_i. For a more precise definition, see the Appendix.

Definition 5.6. A divisor of the form $(G(u, v)) - (H(u, v))$ – that is, the *divisor of the rational function* $G(u, v)/H(u, v)$ – is called a *principal divisor*. We let $\mathbb{J}$ (more precisely, $\mathbb{J}(\mathbb{K})$, where $\mathbb{K}$ is a field containing $\mathbb{F}$) denote the quotient of the group $\mathbb{D}^0$ of divisors of degree zero defined over $\mathbb{K}$ by the subgroup $\mathbb{P}$ of principal divisors coming from $G, H \in \mathbb{K}[u, v]$. $\mathbb{J} = \mathbb{D}^0/\mathbb{P}$ is called the *jacobian* of the curve.

5.2 Addition on the Jacobian

Definitions 5.3 and 5.6 apply to any curve C. Why, then, do we insist on working with the jacobian group of a *hyperelliptic* curve? The first reason is that Definition 5.6 is rather abstract – $\mathbb{J}$ is defined as the quotient of one infinite group by another. In order to set up computations on $\mathbb{J}$ one needs an easily described set of divisors that represent the equivalence classes of $\mathbb{D}^0$ modulo $\mathbb{P}$. In the case of hyperelliptic curves, one can show (either using the Riemann–Roch theorem as in [Fulton 1969] or in a more elementary way as in the Appendix) that every element of $\mathbb{J}$ can be uniquely represented by a so-called *reduced* divisor.

Definition 5.7. A divisor $D = \sum m_i P_i - (*)\infty \in \mathbb{D}^0$ is said to be reduced if:

1) All of the m_i are non-negative, and $m_i \leq 1$ if P_i is equal to its opposite.
2) If $P_i \neq \widetilde{P}_i$, then P_i and $\widetilde{P}_i$ do not both occur in the sum.
3) $\sum m_i \leq g$.

Any reduced divisor $D = \sum m_i P_i - (*)\infty \in \mathbb{D}^0$ can be uniquely represented as the g.c.d. of the divisor of the function $a(u) = \prod(u - x_i)^{m_i}$ and the divisor of the function $b(u) - v$, where $P_i = (x_i, y_i)$ and $b(u)$ is the unique polynomial of degree less than $\deg(a(u))$ such that $b(x_i) = y_i$ for each i and $b(u)^2 + h(u)b(u) - f(u)$ is divisible by $a(u)$. (See the Appendix for a proof.) If D is represented in this way, we write $D = \text{div}(a, b)$.

The second reason why we work with hyperelliptic rather than more general curves is that it is relatively straightforward to add two elements of $\mathbb{J}$. More precisely, given two reduced divisors $D_1 = \text{div}(a_1, b_1)$ and $D_2 = \text{div}(a_2, b_2)$, it is not hard to compute the reduced divisor D_3 that is equivalent to $D_1 + D_2$ in the group $\mathbb{J}$. The algorithm to do this is closely analogous to the classical number-theoretic algorithm for composing two binary quadratic forms. This algorithm goes back to Gauss; for a modern treatment see, for example, Chapters 9–10 of [Rose 1994].

There is a conceptual explanation for the existence of an algorithm for addition on the jacobian of a hyperelliptic curve that is similar to the algorithm for composing quadratic forms. From a modern viewpoint, the equivalence classes of binary quadratic forms are elements of the divisor class group (usually called the *ideal class group*) of the imaginary quadratic field $\mathbb{Q}(\sqrt{d})$ (where d is the discriminant of the quadratic forms). In an analogous way, the hyperelliptic curve (20) gives rise to the *function field* consisting of rational functions $G(u, v)/H(u, v)$ considered modulo the quadratic relation $v^2 + h(u)v = f(u)$. This function field is a quadratic extension of the basic field $\mathbb{K}(u)$, just as $\mathbb{Q}(\sqrt{d})$ is a quadratic extension of the basic field $\mathbb{Q}$. Moreover, the definition of the jacobian – the quotient of the degree 0 divisors by the divisors of rational functions – is analogous to the definition of the ideal class group of $\mathbb{Q}(\sqrt{d})$ as the quotient of the divisors (ideals) by the principal ideals generated by elements of $\mathbb{Q}(\sqrt{d})$.

If our curve C were given by a more complicated equation in u and v, in which v occurred to powers greater than 2, then this analogy would no longer hold, and in most cases it would be much more difficult to compute in the jacobian of the curve.

The algorithm for adding two reduced divisors $D_1 = \text{div}(a_1, b_1)$ and $D_2 = \text{div}(a_2, b_2)$ in $\mathbb{J}$ is described in detail in the Appendix. Here we shall give the algorithm only in the special case when the polynomials $a_1(u)$ and $a_2(u)$ are relatively prime. In that case we can use the Euclidean algorithm for polynomials (see §3 of Chapter 3) to write $s_1 a_1 + s_2 a_2 = 1$ for some polynomials $s_1(u)$ and $s_2(u)$. We set $a = a_1 a_2$, and we let b equal $s_1 a_1 b_2 + s_2 a_2 b_1$ modulo a (that is, b is the remainder when the polynomial $s_1 a_1 b_2 + s_2 a_2 b_1$ is divided by a). If $\deg(a) \leq g$, we are done: we set $D_3 = \text{div}(a, b)$. Otherwise, we set $a' = (f - hb - b^2)/a$ and then $b' = -h - b$ modulo a'. If $\deg(a') \leq g$, we set $D_3 = \text{div}(a', b')$. Otherwise, we set $a'' = (f - hb' - b'^2)/a'$ and $b'' = -h - b'$ modulo a'', and so on, until we obtain $D_3 = \text{div}(a_3, b_3)$ with $\deg(a_3) \leq g$. See the Appendix for a proof that (1) the degree of $a, a', a'', \ldots$ keeps decreasing until it becomes less than or equal to g, and (2) $D_3 = D_1 + D_2$ in $\mathbb{J}$.

5.3 The Zeta-Function

As in the case of elliptic curves, the zeta-function of a hyperelliptic curve is a basic tool in computations. Let $\mathbb{J}$ be the jacobian of a hyperelliptic curve C defined over $\mathbb{F}_q$ and given by the equation $v^2 + h(u)v = f(u)$. Let $\mathbb{F}_{q^r}$ denote the degree-r extension of $\mathbb{F}_q$, and let N_r denote the order of the (finite) abelian group $\mathbb{J}(\mathbb{F}_{q^r})$. Denote by M_r the number of $\mathbb{F}_{q^r}$-points on C, including the point at infinity.

Definition 5.8. Let C be a hyperelliptic curve defined over $\mathbb{F}_q$, and let $M_r = \#C(\mathbb{F}_{q^r})$ for $r \geq 1$. The *zeta-function* of C is the power series

$$Z(C/\mathbb{F}_q; T) = e^{\sum_{r \geq 1} M_r T^r / r} . \tag{21}$$

The following theorem about the zeta-function, generalizing Theorem 1.1, was proved by A. Weil.

Theorem 5.1. *Let C be a hyperelliptic curve of genus g defined over $\mathbb{F}_q$, and let $Z(C/\mathbb{F}_q; T)$ be the zeta-function of C.*

1) $Z(C/\mathbb{F}_q; T)$ is a rational function of the form

$$Z(C/\mathbb{F}_q; T) = \frac{P(T)}{(1 - T)(1 - qT)} , \tag{22}$$

where $P(T)$ is a polynomial of degree $2g$ with integer coefficients of the form

$$P(T) = 1 + a_1 T + \cdots + a_{g-2}T^{g-2} + a_{g-1}T^{g-1} + a_g T^g$$
$$+ qa_{g-1}T^{g+1} + q^2 a_{g-2}T^{g+2} + \cdots + q^{g-1}a_1 T^{2g-1} + q^g T^{2g} \ .$$

2) *$P(T)$ factors as*

$$P(T) = \prod_{i=1}^{g}(1 - \alpha_i T)(1 - \overline{\alpha}_i T) \ ,$$

where each α_i is a complex number of absolute value $\sqrt{q}$, and $\overline{\alpha}_i$ denotes the complex conjugate of α_i.

3) *$N_r = \#\mathbb{J}(\mathbb{F}_{q^r})$ is given by*

$$N_r = \prod_{i=1}^{g}|1 - \alpha_i^r|^2 \ , \tag{23}$$

where $|\ |$ denotes the usual complex absolute value. In particular, $N_1 = P(1)$.

From Theorem 5.1 it follows that to compute N_r for arbitrary r one needs only to (i) find the coefficients $a_1, a_2, \ldots, a_g$ of $P(T)$, hence determining $P(T)$; (ii) factor $P(T)$ to obtain the α_i; and (iii) use equation (23). One can find $a_1, a_2, \ldots, a_g$ by computing $M_1, M_2, \ldots, M_g$. Namely, we note that if we multiply both sides of equation (22) by $(1 - T)(1 - qT)$, we obtain

$$P(T) = (1 - T)(1 - qT)Z(C/\mathbb{F}_q; T) \ .$$

Taking logarithms of both sides, using (21), and then differentiating with respect to T, we have

$$\frac{P'(T)}{P(T)} = \sum_{r \geq 0}(M_{r+1} - 1 - q^{r+1})T^r \ .$$

By equating coefficients of $T^0, T^1, \ldots, T^{g-1}$ on both sides, we see that the first g values $M_1, M_2, \ldots, M_g$ suffice to determine the coefficients $a_1, a_2, \ldots, a_g$, and hence N_r for all r.

The following procedure determines N_r in the case $g = 2$:

1) By exhaustive search, compute M_1 and M_2.
2) The coefficients of $Z(C/\mathbb{F}_q; T)$ are given by

$$a_1 = M_1 - 1 - q \quad \text{and} \quad a_2 = (M_2 - 1 - q^2 + a_1^2)/2 \ .$$

3) Solve the quadratic equation $X^2 + a_1 X + (a_2 - 2q) = 0$ to obtain two solutions γ_1 and γ_2.
4) Solve $X^2 - \gamma_1 X + q = 0$ to obtain a solution α_1, and solve $X^2 - \gamma_2 X + q = 0$ to obtain a solution α_2.
5) Then $N_r = |1 - \alpha_1^r|^2 \cdot |1 - \alpha_2^r|^2$.

Exercises for §5

1. Let $\mathbb{J}$ be the jacobian of a hyperelliptic curve C of genus g defined over $\mathbb{F}_q$. Show that $N_r = \#\mathbb{J}(\mathbb{F}_{q^r})$ lies in the interval $\left[(q^{r/2} - 1)^{2g}, (q^{r/2} + 1)^{2g}\right]$.

2. Suppose that the numerator of $Z(C/\mathbb{F}_q; T)$ factors over $\mathbb{Q}$ into a product of polynomials of degree $\leq d_0$. Prove that any prime divisor of $N_r = \#\mathbb{J}(\mathbb{F}_{q^r})$ is no greater than $(q^{r/2} + 1)^{d_0}$. Thus, we have a better chance of getting groups with non-smooth order (see Definition 2.2) if the numerator of $Z(C/\mathbb{F}_q; T)$ is irreducible over $\mathbb{Q}$. (See [Koblitz 1991c] for results on irreducibility of the numerator of $Z(C/\mathbb{F}_q; T)$ for certain families of curves.)

3. Suppose that C has genus 2 and is defined over $\mathbb{F}_p$, where $p > 2$. If $M_1 \equiv M_2 \equiv 1 \pmod{p}$, prove that $N_r \equiv 1 \pmod{p}$ for all r.

4. Let C be a hyperelliptic curve of the form $v^2 + v = f(u)$ defined over $\mathbb{F}_2$. Prove that $M_1 \equiv 1 \pmod{2}$ and $M_2 \equiv 1 \pmod{4}$. If C has genus 2, prove that N_r is odd for all r.

5. Let C have the form $v^2 + uv = f(u)$ over $\mathbb{F}_2$. Prove that $M_1 \equiv 0 \pmod{2}$ and $M_2 \equiv 0 \pmod{4}$. If $g = 2$, prove that N_r is even for all r.

6. Let $g = 2$. By analogy with Exercise 15 of §1, find a recursive relation involving γ_1 and γ_2 that makes it possible to compute the sequence N_r very rapidly.

§6. Hyperelliptic Cryptosystems

The elliptic curve Diffie–Hellman key exchange and ElGamal message transmission that we discussed in §2 carry over word for word to the jacobian group of a hyperelliptic curve.

To implement a hyperelliptic discrete log cryptosystem, a suitable curve C and underlying finite field $\mathbb{F}_q$ must be selected. It is crucial that the order $\#\mathbb{J}(\mathbb{F}_q)$ of the jacobian of C be divisible by a large prime number (see §2.3). Given the current state of computer technology, $\#\mathbb{J}(\mathbb{F}_q)$ should be divisible by a prime number l of at least 40 decimal digits. In addition, to avoid the attack of Frey[*] and Rück [1994], which, generalizing [Menezes, Okamoto, and Vanstone 1993], reduces the discrete logarithm problem in $\mathbb{J}(\mathbb{F}_q)$ to the discrete logarithm problem in an extension field $\mathbb{F}_{q^k}^*$, l should not divide $q^k - 1$ for any small k (say, $1 \leq k \leq 2000/(\log_2 q)$).

A secondary consideration is that we would like for there to be efficient implementations of the arithmetic in $\mathbb{F}_q$; finite fields of characteristic 2 appear to be the most attractice from this point of view.

---

[*] This is the same Frey who in 1985 had the idea that ultimately led to Wiles' proof of Fermat's Last Theorem (see §4.2); he subsequently became interested in elliptic and hyperelliptic cryptography.

6.1 Examples in Characteristic 2

Example 6.1. Consider the following hyperelliptic curve C of genus 2 defined over $\mathbb{F}_2$:

$$C \ : \ v^2 + v = u^5 + u^3 + u \ .$$

By a simple count we find that $M_1 = 3$ and $M_2 = 9$; hence $a_1 = 0$ and $a_2 = 2$. The solutions of $X^2 - 2 = 0$ are $\gamma_1 = \sqrt{2}$ and $\gamma_2 = -\sqrt{2}$. Solving $X^2 - \sqrt{2}X + 2 = 0$, we have $\alpha_1 = (\sqrt{2} + \sqrt{6}i)/2$; and solving $X^2 + \sqrt{2}X + 2 = 0$, we obtain $\alpha_2 = (-\sqrt{2} + \sqrt{6}i)/2$. The numerator of the zeta-function is $1 + 2T^2 + 4T^4$, and

$$N_r = |1 - \alpha_1^r|^2 \cdot |1 - \alpha_2^r|^2 = \begin{cases} 2^{2r} + 2^r + 1 \ , & \text{if } r \equiv 1, 5 \bmod 6 \ , \\ (2^r + 2^{r/2} + 1)^2 \ , & \text{if } r \equiv 2, 4 \bmod 6 \ , \\ (2^r - 1)^2 \ , & \text{if } r \equiv 3 \bmod 6 \ , \\ (2^{r/2} - 1)^4 \ , & \text{if } r \equiv 0 \bmod 6 \ . \end{cases}$$

For $r = 101$,

$N_{101} =$

$= 6427752177035961102167848369367185711289268433934164747616257,$

which has prime factorization

$7 \cdot 607 \cdot 1512768222413735255864403005264105839324374778520631853993.$

Hence, N_{101} is divisible by a 58-decimal digit prime l_{58}. However, since l_{58} divides $(2^{101})^3 - 1$, the system is vulnerable to the Frey–Rück attack, and offers us no more security than a discrete log system in $\mathbb{F}_{2^{303}}$. Hence the curve C is not suitable for cryptographic applications.

Example 6.2. The hyperelliptic curve $v^2 + v = u^{383} + 1$ over $\mathbb{F}_2$ has genus $g = 191$. Because of the special form of this curve, its zeta-function can easily be computed (see [Koblitz 1991c]); its numerator is $1 + (711 \cdot 2^{87})T^{191} + 2^{191}T^{382}$. It turns out that $N = N_1$ is of the form $3l_{58}$, where l_{58} is the 58-digit prime

$1046183622564446793972631570497937095686563183433452530347 \ .$

Unlike Example 6.1, this curve is not vulnerable to the Frey–Rück reduction, since $l_{58} \nmid 2^k - 1$ for $k \leq 2000$. However, there is another reason why it would not be wise to use a hyperelliptic curve of high genus over a small finite field. In [Adleman, DeMarrais, and Huang 1994] the authors give an $L_{p^{2g+1}}(1/2)$-algorithm (see Definition 3.2 of Chapter 2) for the discrete logarithm problem on the jacobian of a high-genus hyperelliptic curve defined over $\mathbb{F}_p$. In other words, for fixed p their algorithm runs in time $\exp(O(\sqrt{g \ln g}))$ for large g. Although the algorithm in its current form would not quite be feasible for the curve in this example, it is close enough to the practical range that one cannot have confidence in the cryptographic suitability of the curve. (Actually, the algorithm of Adleman, DeMarrais, and Huang applies only to odd p; however, it is very likely that an analogous algorithm can be developed for $p = 2$.)

Example 6.3. The hyperelliptic curve $v^2 + v = u^7$ over $\mathbb{F}_2$ has genus $g = 3$. The numerator of its zeta-function (see [Koblitz 1991c]) is $1 - 2T^3 + 8T^6$. It turns out that N_{47} is equal to $7l_{42}$, where l_{42} is the 42-digit prime

$$398227592830903984669824190479460780961207 \ .$$

Neither the Frey–Rück reduction nor the Adleman–DeMarrais–Huang algorithm can feasibly be applied to this example.

Example 6.4. (a) Consider the curve $v^2 + uv = u^5 + u^2 + 1$ over $\mathbb{F}_2$. The numerator of the zeta-function (see [Koblitz 1989]) is $1 - T - 2T^3 + 4T^4$. One finds that N_{61} is equal to $2l_{37}$, where l_{37} is the 37-digit prime

$$2658455988447243530986550320280662477 \ .$$

(b) If C is the curve $v^2 + uv = u^5 + 1$ over $\mathbb{F}_2$, then the numerator of $Z(C/\mathbb{F}_2; T)$ is $1 + T + 2T^3 + 4T^4$, and one finds that N_{67} is equal to $8l_{40}$, where l_{40} is the 40-digit prime

$$2722258935596872912437464397871092846187 \ .$$

As in Example 6.3, there is no known subexponential time algorithm that can feasibly be applied to these two examples.

6.2 Example over a Large Prime Field

Example 6.5. Let $n = 2g + 1$ be an odd prime, and let $p \equiv 1 \pmod{n}$. Consider the hyperelliptic curve

$$C : \quad v^2 + v = u^n \tag{24}$$

over $\mathbb{F}_p$. Its jacobian $\mathbb{J}$ is a quotient of the jacobian of the famous Fermat curve $X^n + Y^n = 1$, which in characteristic zero has no nontrivial rational points by Fermat's Last Theorem ([Wiles 1995] and [Taylor and Wiles 1995]). These jacobians have been studied for many years; in fact, it was the zeta-functions of Fermat curves and "diagonal" hypersurfaces that André Weil cited as evidence for his famous conjectures [Weil 1949]. A detailed treatment can be found, for example, in [Ireland and Rosen 1990]. I will state what we need without proof.

Let $\zeta = e^{2\pi i/n}$, and let $\alpha \in \mathbb{F}_p$ be a fixed non-nth-power. There is a unique multiplicative map χ on $\mathbb{F}_p^*$ such that $\chi(\alpha) = \zeta$. We extend this character χ to $\mathbb{F}_p$ by setting $\chi(0) = 0$. The Jacobi sum of the character χ with itself is defined as follows:

$$J(\chi, \chi) = \sum_{y \in \mathbb{F}_p} \chi(y)\chi(1 - y) \ . \tag{25}$$

For $1 \le i \le n-1$ let σ_i be the automorphism of the field $\mathbb{Q}(\zeta)$ such that $\sigma_i(\zeta) = \zeta^i$. Then an easy counting argument shows that the number of points on the curve (24), including the point at infinity, is equal to

$$M = p + 1 + \sum_{i=1}^{n-1} \sigma_i(J(\chi, \chi)) \; ;$$

and one can also show (see [Weil 1949] and [Ireland and Rosen 1990]) that $-J(\chi, \chi)$ and its conjugates are the reciprocal roots of the numerator of the zeta-function of this curve. In other words,

$$Z(C/\mathbb{F}_p; T) = \frac{\prod_{i=1}^{n-1}\left(1 + \sigma_i(J(\chi, \chi))T\right)}{(1-T)(1-pT)} .$$

The number N of points on the jacobian $\mathbb{J}$ of C is equal to the value at 1 of the numerator of $Z(C/\mathbb{F}_p; T)$; that is,

$$N = \prod_{i=1}^{n-1} \sigma_i(J(\chi, \chi) + 1) = \mathbb{N}(J(\chi, \chi) + 1) , \tag{26}$$

where $\mathbb{N}$ denotes the norm of an algebraic number.

Along with the curve C given by (24), we also consider its "twists" by non-nth-powers and by non-squares. To do this, let β be a fixed non-square in $\mathbb{F}_p$, and consider the equation

$$\beta^{-i}\left(v + \tfrac{1}{2}\right)^2 = \alpha^j u^n + \tfrac{1}{4}$$

for $i = 0, 1$ and $j = 0, 1, \ldots, n - 1$, where α is the fixed non-nth-power that was chosen above. This equation can be rewritten in the form

$$v^2 + v + (1 - \beta^i)/4 = \beta^i \alpha^j u^n . \tag{27}$$

By analogy with (26) one finds that the number of points on the jacobian of the curve (27) is given by

$$N_{i,j} = \mathbb{N}\left(J(\chi, \chi) + (-1)^i \zeta^j\right) , \qquad i = 0, 1 , \quad j = 0, 1, \ldots, n - 1 . \tag{28}$$

When $i = 0$, it follows from (30) below that $N_{0,0}$ is divisible by n^2 and $N_{0,j}$ is divisible by n for $j = 1, 2, \ldots, n - 1$. Hence, in that case the most one can hope for is that $N_{0,0}/n^2$ or $N_{0,j}/n$ be a prime. When $i = 1$, there is no such obstruction to $N_{1,j}$ itself being prime. Thus, after we compute $J(\chi, \chi)$ for our chosen n and $p \equiv 1 \pmod{n}$, we will want to compute the numbers (28) and test $n^{-2}N_{0,0}$, $n^{-1}N_{0,j}$, and $N_{1,j}$ for primality, $j = 0, 1, \ldots, n - 1$. Since $N_{i,j}$ is of order $p^g = p^{(n-1)/2}$, we see that to get jacobians whose order is divisible by a prime of at least 40 digits we should choose p greater than the bounds in the following table:

n	3	5	7	11	13	17
p	$> 10^{40}$	$> 10^{20}$	$> 2 \times 10^{13}$	$> 10^8$	$> 5 \times 10^6$	$> 10^5$

First suppose that $n \geq 13$. Since p has order of magnitude 5000000 for $n = 13$ and less for $n > 13$, it is feasible to compute $J(\chi, \chi)$ from the definition (25). However, because of the Adleman–DeMarrais–Huang algorithm, one should choose n and p so that $\ln p > n$; so we probably should take $n = 13$. Thus, we might choose $n = 13$ and $p > 5000000$, $p \equiv 1 \pmod{13}$. For each such p, we compute $J(\chi, \chi)$ from (25) and test the 26 numbers $\frac{1}{169} N_{0,0}$, $\frac{1}{13} N_{0,j}$, and $N_{1,j}$ for primality. When we find that the number in (28) is a prime or 13 or 169 times a prime, we can use the corresponding equation (27) for our hyperelliptic cryptosystem.

For $n \leq 11$ we would like to have a way of computing the Jacobi sum $J(\chi, \chi)$ that is much faster than the definition (25). In the case $n = 3$ – that is, when C is an elliptic curve – this is easy (and essentially goes back to Gauss, who gave an explicit formula for the number of points on the curve (24) when $n = 3$). Namely, let a be an integer such that $a^3 \equiv 1 \pmod{p}$, $a \not\equiv 1 \pmod{p}$; such an a can be found by computing $\alpha^{(p-1)/3}$ in $\mathbb{F}_p$. (Recall that α is a fixed non-cube in $\mathbb{F}_p$.) Now compute the greatest common divisor of p and $a - \zeta$ in the Euclidean ring $\mathbb{Z}[\zeta]$, $\zeta = (-1 + i\sqrt{3})/2$. Then $J(\chi, \chi) = \pm\zeta^k \text{g.c.d.}(p, a - \zeta)$, where the root of unity $\pm\zeta^k$ is chosen so that $J(\chi, \chi) \equiv -1 \pmod{3}$ in $\mathbb{Z}[\zeta]$. This method of computing $J(\chi, \chi)$ is very fast, taking $O(\ln^3 p)$ bit operations.

For $n = 2g + 1 \geq 5$, if we choose p to be a generalized Mersenne prime of the form

$$p = \frac{a^n - 1}{a - 1} ,$$

then it is again very easy to compute $J(\chi, \chi)$. Suppose that α, our fixed non-nth-power modulo p, is chosen so that $\alpha^{(p-1)/n} \equiv a \pmod{p}$. Then one can show that

$$J(\chi, \chi) = \pm\zeta^k \prod_{i=1}^{g} (a - \sigma_i^{-1}(\zeta)) , \tag{29}$$

where $\pm\zeta^k$ is chosen so that

$$J(\chi, \chi) \equiv -1 \pmod{(\zeta - 1)^2} \tag{30}$$

in the ring $\mathbb{Z}[\zeta]$ (see p. 227 of [Ireland and Rosen 1990] for (30)). Using the fact that

$$\zeta^j = (1 + \zeta - 1)^j \equiv 1 + j(\zeta - 1) \pmod{(\zeta - 1)^2} ,$$

it is trivial to find the value $\pm\zeta^k$, which depends only on a modulo n. In the case $n = 5$, this root of unity is given in the following table:

$a \bmod 5$	0	2	3	4	
$\pm\zeta^k$	$-\zeta$	$-\zeta^4$	ζ^2	ζ^3	(31)

Letting $n = 5$ and $a \geq 10^5$, I quickly found the following cases where $p = a^4 + a^3 + a^2 + a + 1$ is prime and the order of the jacobian over $\mathbb{F}_p$ is divisible by a large prime:

$$a = 100003 , \qquad p = 100013000640014200121 ,$$

$$N_{0,1} = 5 \cdot 20005200592038621583241900701806833302981 ;$$

$$a = 100012 , \qquad p = 100049009010736922621 ,$$

$$N_{0,4} = 5 \cdot 20019608400055154071899804430461276558801 ;$$

and

$$a = 100018 \ , \qquad p = 100073019992433811151 \ ,$$

$$N_{1,0} = 10014609331407177786767800456957577013341 \ .$$

In the general case when p is not of the form $(a^n - 1)/(a - 1)$, it is still possible to determine $J(\chi, \chi)$ in polynomial time in $\ln p$ for fixed n. The key step is to find a single generator for a prime ideal of $\mathbb{Z}[\zeta]$ lying over p; one way to do this is to use the LLL-algorithm [Lenstra, Lenstra, and Lovász 1982] to find a short vector in the lattice in $\mathbb{R}^{n-1}$ corresponding to the ideal generated by the two elements p and $(a - \zeta)$ (where, as before, the rational integer a is an n-th root of unity modulo p). More details can be found in [Buhler and Koblitz 1998]; see also [Lenstra 1975] and [Buchmann and Williams 1987].

6.3 Future Work

There are several areas of research that need to be pursued before hyperelliptic curve cryptosystems are adopted in practical applications.

1) As in the case of elliptic curve cryptosystems, a key security question is whether there exists a subexponential time algorithm for the discrete log problem in the general case or for special classes of curves.

2) It would be worthwhile to investigate the conditions under which the reduction in [Frey and Rück 1994] leads to a subexponential time algorithm. Most likely, except in the "supersingular case" (when all of the reciprocal roots of the zeta-function have the same "p-adic norm", as in Example 6.1 but not in Examples 6.2–6.5), this almost never occurs.

3) The algorithm in [Adleman, DeMarrais, and Huang 1994] should be improved upon and extended to the case $p = 2$ and the case of powers $q = p^f$ with $f > 1$.

4) Further research needs to be done on the efficient implementation of the addition rule in the jacobian. Slightly more efficient algorithms may arise if one considers different forms of the defining equation. Some asymptotically faster variants of the reduction algorithm in the Appendix are described by Cantor [1987] (for large g) and by Petersen [1994] (in the case $g = 2$).

5) One of the methods of looking for a suitable hyperelliptic curve is to select at random a defining equation over a large finite field $\mathbb{F}_q$ and compute $\#J(\mathbb{F}_q)$ directly. Pila [1990] presented a generalization of the algorithm in [Schoof 1985] that does this in deterministic polynomial time (for fixed genus). As has already happened in the case of elliptic curves, further work is likely to lead to simplifications and increased efficiency, so that it becomes feasible to compute the order of random jacobian groups. (See also [Poonen 1996] and [Adleman and Huang 1996].)

Exercises for § 6

1. Using (29) and (31), find $J(\chi, \chi)$ when $n = 5$, $p = 11$.

2. In Example 6.5, make a table of $\pm \zeta^k$ when $n = 7$ (see (31)).

Appendix. An Elementary Introduction to Hyperelliptic Curves

by Alfred J. Menezes, Yi-Hong Wu, and Robert J. Zuccherato

This appendix is an elementary introduction to some of the theory of hyperelliptic curves over finite fields of arbitrary characteristic that has cryptographic relevance. Cantor's algorithm for adding in the jacobian of a hyperelliptic curve is presented, along with a proof of its correctness.

Hyperelliptic curves are a special class of algebraic curves and can be viewed as generalizations of elliptic curves. There are hyperelliptic curves of every genus $g \geq 1$. A hyperelliptic curve of genus $g = 1$ is an elliptic curve. Elliptic curves have been extensively studied for over a hundred years, and there are many books on the topic (for example, [Silverman 1986 and 1994], [Husemöller 1987], [Koblitz 1993], [Menezes 1993]).

On the other hand, the theory of hyperelliptic curves has not received as much attention by the research community. Most results concerning hyperelliptic curves which appear in the literature on algebraic geometry are couched in very general terms. For example, a common source cited in papers on hyperelliptic curves is [Mumford 1984]. However, the non-specialist will have difficulty specializing (not to mention finding) the results in this book to the particular case of hyperelliptic curves. Another difficulty one encounters is that the theory in such books is usually restricted to the case of hyperelliptic curves over the complex numbers (as in Mumford's book), or over algebraically closed fields of characteristic not equal to 2. The recent book [Cassels and Flynn 1996] is an extensive account of curves of genus 2. (Compared to their book, our approach is definitely "low-brow".)

Recently, applications of hyperelliptic curves have been found in areas outside algebraic geometry. Hyperelliptic curves were a key ingredient in Adleman and Huang's random polynomial-time algorithm for primality proving [Adleman and Huang 1992]. Hyperelliptic curves have also been considered in the design of error-correcting codes [Brigand 1991], in the evaluation of definite integrals [Bertrand 1995], in integer factorization algorithms [Lenstra, Pila and Pomerance 1993], and in public-key cryptography (see Chapter 6 of the present book). Hyperelliptic

Authors' addresses: Alfred J. Menezes, Department of Combinatorics and Optimization, University of Waterloo, Waterloo, Ontario, Canada N1L 3G1, e-mail: ajmeneze@math. uwaterloo.ca; Yi-Hong Wu, Department of Discrete and Statistical Sciences, Auburn University, Auburn, AL 36849, USA; Robert J. Zuccherato, Entrust Technologies, 750 Heron Road, Ottawa, Ontario, Canada K1V 1A7, e-mail: robert.zuccherato@entrust.com.

curves over finite fields of characteristic two are particularly of interest when implementing codes and cryptosystems.

Charlap and Robbins [1988] presented an elementary introduction to elliptic curves. The purpose was to provide elementary self-contained proofs of some of the basic theory relevant to Schoof's algorithm [Schoof 1985] for counting the points on an elliptic curve over a finite field. The discussion was restricted to fields of characteristic not equal to 2 or 3. However, for practical applications, elliptic and hyperelliptic curves over characteristic two fields are especially attractive. This appendix, similar in spirit to the paper of Charlap and Robbins, presents an elementary introduction to some of the theory of hyperelliptic curves over finite fields of arbitrary characteristic. For a general introduction to the theory of algebraic curves, consult [Fulton 1969].

§ 1. Basic Definitions and Properties

Definition 1.1. Let $\mathbb{F}$ be a field and let $\overline{\mathbb{F}}$ be the algebraic closure of $\mathbb{F}$ (see Definition 1.8 of Chapter 3). A *hyperelliptic curve* C *of genus* g *over* $\mathbb{F}$ $(g \geq 1)$ is an equation of the form

$$C \ : \ v^2 + h(u)v = f(u) \qquad \text{in} \qquad \mathbb{F}[u, v] \ , \tag{1}$$

where $h(u) \in \mathbb{F}[u]$ is a polynomial of degree at most g, $f(u) \in \mathbb{F}[u]$ is a monic polynomial of degree $2g + 1$, and there are no solutions $(u, v) \in \overline{\mathbb{F}} \times \overline{\mathbb{F}}$ which simultaneously satisfy the equation $v^2 + h(u)v = f(u)$ and the partial derivative equations $2v + h(u) = 0$ and $h'(u)v - f'(u) = 0$.

A *singular point* on C is a solution $(u, v) \in \overline{\mathbb{F}} \times \overline{\mathbb{F}}$ which simultaneously satisfies the equation $v^2 + h(u)v = f(u)$ and the partial derivative equations $2v + h(u) = 0$ and $h'(u)v - f'(u) = 0$. Definition 1.1 thus says that a hyperelliptic curve does not have any singular points.

For the remainder of this paper it is assumed that the field $\mathbb{F}$ and the curve C have been fixed.

Lemma 1.1. *Let C be a hyperelliptic curve over $\mathbb{F}$ defined by equation (1).*

1) *If $h(u) = 0$, then char$(\mathbb{F}) \neq 2$.*
2) *If char$(\mathbb{F}) \neq 2$, then the change of variables $u \to u$, $v \to (v - h(u)/2)$ transforms C to the form $v^2 = f(u)$ where $\deg_u f = 2g + 1$.*
3) *Let C be an equation of the form (1) with $h(u) = 0$ and char$(\mathbb{F}) \neq 2$. Then C is a hyperelliptic curve if and only if $f(u)$ has no repeated roots in $\overline{\mathbb{F}}$.*

Proof.
1) Suppose that $h(u) = 0$ and char$(\mathbb{F}) = 2$. Then the partial derivative equations reduce to $f'(u) = 0$. Note that $\deg_u f'(u) = 2g$. Let $x \in \overline{\mathbb{F}}$ be a root of the equation $f'(u) = 0$, and let $y \in \overline{\mathbb{F}}$ be a root of the equation $v^2 = f(x)$. Then the point (x, y) is a singular point on C. Statement 1) now follows.

2) Under this change of variables, the equation (1) is transformed to

$$(v - h(u)/2)^2 + h(u)(v - h(u)/2) = f(u) ,$$

which simplifies to $v^2 = f(u) + h(u)^2/4$; note that $\deg_u(f + h^2/4) = 2g + 1$.
3) A singular point (x, y) on C must satisfy $y^2 = f(x)$, $2y = 0$, and $f'(x) = 0$.
Hence $y = 0$ and x is a repeated root of the polynomial $f(u)$. $\square$

Definition 1.2. Let $\mathbb{K}$ be an extension field of $\mathbb{F}$. The *set of* $\mathbb{K}$-*rational points on*
C, denoted $C(\mathbb{K})$, is the set of all points $P = (x, y) \in \mathbb{K} \times \mathbb{K}$ that satisfy the
equation (1) of the curve C, together with a special *point at infinity** denoted ∞.
The set of points $C(\overline{\mathbb{F}})$ will simply be denoted by C. The points in C other than
∞ are called *finite points*.

Example 1.1. The illustrations on the next page show two examples of hyperelliptic
curves over the field of real numbers. Each curve has genus $g = 2$ and $h(u) = 0$.

Definition 1.3. Let $P = (x, y)$ be a finite point on a hyperelliptic curve C. The
opposite of P is the point $\tilde{P} = (x, -y - h(x))$. (Note that $\tilde{P}$ is indeed on C.) We
also define the opposite of ∞ to be $\tilde{\infty} = \infty$ itself. If a finite point P satisfies
$P = \tilde{P}$, then the point is said to be *special*; otherwise, the point is said to be
ordinary.

Example 1.2. Consider the curve $C : v^2 + uv = u^5 + 5u^4 + 6u^2 + u + 3$ over the
finite field $\mathbb{F}_7$. Here, $h(u) = u$, $f(u) = u^5 + 5u^4 + 6u^2 + u + 3$ and $g = 2$. It can
be verified that C has no singular points (other than ∞), and hence C is indeed a
hyperelliptic curve. The $\mathbb{F}_7$-rational points on C are

$$C(\mathbb{F}_7) = \{\infty, (1, 1), (1, 5), (2, 2), (2, 3), (5, 3), (5, 6), (6, 4)\} .$$

The point $(6, 4)$ is a special point.

1) $C_1 : v^2 = u^5 + u^4 + 4u^3 + 4u^2 + 3u + 3 = (u + 1)(u^2 + 1)(u^2 + 3)$. The graph of
C_1 in the real plane is shown below.

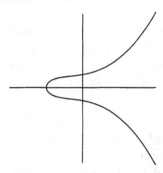

---

* The point at infinity lies in the projective plane $\mathbf{P}^2(\mathbb{F})$. It is the only projective point
lying on the line at infinity that satisfies the homogenized hyperelliptic curve equation. If
$g \geq 2$, then ∞ is a singular (projective) point; this is allowed, since $\infty \notin \overline{\mathbb{F}} \times \overline{\mathbb{F}}$.

2) $C_2 : v^2 = u^5 - 5u^3 + 4u = u(u - 1)(u + 1)(u - 2)(u + 2)$. The graph of C_2 in the real plane is shown below.

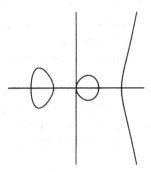

Example 1.3. Consider the finite field $\mathbb{F}_{2^5} = \mathbb{F}_2[x]/(x^5 + x^2 + 1)$, and let α be a root of the primitive polynomial $x^5 + x^2 + 1$ in $\mathbb{F}_{2^5}$. The powers of α are listed in Table 1.

n	α^n	n	α^n	n	α^n
0	1	11	$\alpha^2 + \alpha + 1$	22	$\alpha^4 + \alpha^2 + 1$
1	α	12	$\alpha^3 + \alpha^2 + \alpha$	23	$\alpha^3 + \alpha^2 + \alpha + 1$
2	α^2	13	$\alpha^4 + \alpha^3 + \alpha^2$	24	$\alpha^4 + \alpha^3 + \alpha^2 + \alpha$
3	α^3	14	$\alpha^4 + \alpha^3 + \alpha^2 + 1$	25	$\alpha^4 + \alpha^3 + 1$
4	α^4	15	$\alpha^4 + \alpha^3 + \alpha^2 + \alpha + 1$	26	$\alpha^4 + \alpha^2 + \alpha + 1$
5	$\alpha^2 + 1$	16	$\alpha^4 + \alpha^3 + \alpha + 1$	27	$\alpha^3 + \alpha + 1$
6	$\alpha^3 + \alpha$	17	$\alpha^4 + \alpha + 1$	28	$\alpha^4 + \alpha^2 + \alpha$
7	$\alpha^4 + \alpha^2$	18	$\alpha + 1$	29	$\alpha^3 + 1$
8	$\alpha^3 + \alpha^2 + 1$	19	$\alpha^2 + \alpha$	30	$\alpha^4 + \alpha$
9	$\alpha^4 + \alpha^3 + \alpha$	20	$\alpha^3 + \alpha^2$	31	1
10	$\alpha^4 + 1$	21	$\alpha^4 + \alpha^3$		

Table 1. Powers of α in the finite field $\mathbb{F}_{2^5} = \mathbb{F}_2[x]/(x^5 + x^2 + 1)$

Consider the curve $C : v^2 + (u^2 + u)v = u^5 + u^3 + 1$ of genus $g = 2$ over the finite field $\mathbb{F}_{2^5}$. Here, $h(u) = u^2 + u$ and $f(u) = u^5 + u^3 + 1$. It can be verified that C has no singular points (other than ∞), and hence C is indeed a hyperelliptic curve. The finite points in $C(\mathbb{F}_{2^5})$, the set of $\mathbb{F}_{2^5}$-rational points on C, are:

$$(0, 1) \quad (1, 1) \quad (\alpha^5, \alpha^{15}) \quad (\alpha^5, \alpha^{27}) \quad (\alpha^7, \alpha^4) \quad (\alpha^7, \alpha^{25})$$
$$(\alpha^9, \alpha^{27}) \quad (\alpha^9, \alpha^{30}) \quad (\alpha^{10}, \alpha^{23}) \quad (\alpha^{10}, \alpha^{30}) \quad (\alpha^{14}, \alpha^8) \quad (\alpha^{14}, \alpha^{19})$$
$$(\alpha^{15}, 0) \quad (\alpha^{15}, \alpha^8) \quad (\alpha^{18}, \alpha^{23}) \quad (\alpha^{18}, \alpha^{29}) \quad (\alpha^{19}, \alpha^2) \quad (\alpha^{19}, \alpha^{28})$$
$$(\alpha^{20}, \alpha^{15}) \quad (\alpha^{20}, \alpha^{29}) \quad (\alpha^{23}, 0) \quad (\alpha^{23}, \alpha^4) \quad (\alpha^{25}, \alpha) \quad (\alpha^{25}, \alpha^{14})$$
$$(\alpha^{27}, 0) \quad (\alpha^{27}, \alpha^2) \quad (\alpha^{28}, \alpha^7) \quad (\alpha^{28}, \alpha^{16}) \quad (\alpha^{29}, 0) \quad (\alpha^{29}, \alpha)$$
$$(\alpha^{30}, 0) \quad (\alpha^{30}, \alpha^{16})$$

Of these, the points $(0, 1)$ and $(1, 1)$ are special.

§ 2. Polynomial and Rational Functions

This section introduces basic properties of polynomials and rational functions that arise when they are viewed as functions on a hyperelliptic curve.

Definition 2.1. The *coordinate ring of C over* $\mathbb{F}$, denoted $\mathbb{F}[C]$, is the quotient ring
$$\mathbb{F}[C] = \mathbb{F}[u, v]/(v^2 + h(u)v - f(u)) ,$$
where $(v^2 + h(u)v - f(u))$ denotes the ideal in $\mathbb{F}[u, v]$ generated by the polynomial $v^2 + h(u)v - f(u)$.(See Example 4.1 in Chapter 3 for the definition of "quotient ring".) Similarly, the *coordinate ring of C over* $\overline{\mathbb{F}}$ is defined as
$$\overline{\mathbb{F}}[C] = \overline{\mathbb{F}}[u, v]/(v^2 + h(u)v - f(u)) .$$

An element of $\overline{\mathbb{F}}[C]$ is called a *polynomial function* on C.

Lemma 2.1. *The polynomial* $r(u, v) = v^2 + h(u)v - f(u)$ *is irreducible over* $\overline{\mathbb{F}}$, *and hence* $\overline{\mathbb{F}}[C]$ *is an integral domain.*

Proof. If $r(u, v)$ were reducible over $\overline{\mathbb{F}}$, it would factor as $(v - a(u))(v - b(u))$ for some $a, b \in \overline{\mathbb{F}}[u]$. But then $\deg_u(a \cdot b) = \deg_u f = 2g + 1$ and $\deg_u(a + b) = \deg_u h \leq g$, which is impossible. □

Observe that for each polynomial function $G(u, v) \in \overline{\mathbb{F}}[C]$, we can repeatedly replace any occurrence of v^2 by $f(u) - h(u)v$, so as to eventually obtain a representation
$$G(u, v) = a(u) - b(u)v , \quad \text{where } a(u), b(u) \in \overline{\mathbb{F}}[u] .$$

It is easy to see that the representation of $G(u, v)$ in this form is unique.

Definition 2.2. Let $G(u, v) = a(u) - b(u)v$ be a polynomial function in $\overline{\mathbb{F}}[C]$. The *conjugate* of $G(u, v)$ is defined to be the polynomial function $\overline{G}(u, v) = a(u) + b(u)(h(u) + v)$.

Definition 2.3 Let $G(u, v) = a(u) - b(u)v$ be a polynomial function in $\overline{\mathbb{F}}[C]$. The *norm* of G is the polynomial function $N(G) = G\overline{G}$.

The norm function will be useful in transforming questions about polynomial functions in two variables into easier questions about polynomials in a single variable.

Lemma 2.2. *Let* $G, H \in \overline{\mathbb{F}}[C]$ *be polynomial functions.*
1) $N(G)$ *is a polynomial in* $\overline{\mathbb{F}}[u]$.
2) $N(\overline{G}) = N(G)$.
3) $N(GH) = N(G)N(H)$.

Proof. Let $G = a - bv$ and $H = c - dv$, where $a, b, c, d \in \overline{\mathbb{F}}[u]$.[*]
1) Now, $\overline{G} = a + b(h + v)$ and

$$N(G) = G \cdot \overline{G} = (a - bv)(a + b(h + v)) = a^2 + abh - b^2 f \in \overline{\mathbb{F}}[u] \ .$$

2) The conjugate of $\overline{G}$ is

$$\overline{\overline{G}} = (a + bh) + (-b)(h + v) = a - bv = G \ .$$

Hence $N(\overline{G}) = \overline{G}\,\overline{\overline{G}} = \overline{G}G = N(G)$.
3) $GH = (ac + bdf) - (bc + ad + bdh)v$, and its conjugate is

$$\overline{GH} = (ac + bdf) + (bc + ad + bdh)(h + v)$$
$$= ac + bdf + bch + adh + bdh^2 + bcv + adv + bdhv$$
$$= ac + bc(h + v) + ad(h + v) + bd(h^2 + hv + f)$$
$$= ac + bc(h + v) + ad(h + v) + bd(h^2 + 2hv + v^2)$$
$$= (a + b(h + v))(c + d(h + v))$$
$$= \overline{G}\,\overline{H} \ .$$

Hence $N(GH) = GH\overline{GH} = GH\overline{G}\,\overline{H} = G\overline{G}H\overline{H} = N(G)N(H)$. $\square$

Definition 2.4.   The *function field* $\mathbb{F}(C)$ *of* C *over* $\mathbb{F}$ is the field of fractions of $\mathbb{F}[C]$. Similarly, the *function field* $\overline{\mathbb{F}}(C)$ *of* C *over* $\overline{\mathbb{F}}$ is the field of fractions of $\overline{\mathbb{F}}[C]$. The elements of $\overline{\mathbb{F}}(C)$ are called *rational functions* on C.

Note that $\overline{\mathbb{F}}[C]$ is a subring of $\overline{\mathbb{F}}(C)$, i.e., every polynomial function is also a rational function.

Definition 2.5. Let $R \in \overline{\mathbb{F}}(C)$, and let $P \in C$, $P \neq \infty$. Then R is said to be *defined at* P if there exist polynomial functions $G, H \in \overline{\mathbb{F}}[C]$ such that $R = G/H$ and $H(P) \neq 0$; if no such $G, H \in \overline{\mathbb{F}}[C]$ exist, then R is *not defined* at P. If R is defined at P, the *value of* R *at* P is defined to be $R(P) = G(P)/H(P)$.

It is easy to see that the value $R(P)$ is well-defined, i.e., it does not depend on the choice of G and H. The following definition introduces the notion of the degree of a polynomial function.

Definition 2.6. Let $G(u, v) = a(u) - b(u)v$ be a nonzero polynomial function in $\overline{\mathbb{F}}[C]$. The *degree* of G is defined to be

$$\deg(G) = \max\{2 \deg_u(a),\ 2g + 1 + 2 \deg_u(b)\} \ .$$

Lemma 2.3. *Let* $G, H \in \overline{\mathbb{F}}[C]$.
1) $\deg(G) = \deg_u(N(G))$.

---

[*] If not explicitly stated otherwise, the variable in all polynomials will henceforth be assumed to be u.

2) $\deg(GH) = \deg(G) + \deg(H)$.

3) $\deg(G) = \deg(\overline{G})$.

Proof.

1) Let $G = a(u) - b(u)v$. The norm of G is $N(G) = a^2 + abh - b^2 f$. Let $d_1 = \deg_u(a(u))$ and $d_2 = \deg_u(b(u))$. By the definition of a hyperelliptic curve, $\deg_u(h(u)) \leq g$ and $\deg_u(f(u)) = 2g + 1$. There are two cases to consider:

Case 1: If $2d_1 > 2g + 1 + 2d_2$ then $2d_1 \geq 2g + 2 + 2d_2$, and hence $d_1 \geq g + 1 + d_2$. Hence

$$\deg_u(a^2) = 2d_1 \geq d_1 + g + 1 + d_2 > d_1 + d_2 + g \geq \deg_u(abh) \ .$$

Case 2: If $2d_1 < 2g + 1 + 2d_2$ then $2d_1 \leq 2g + 2d_2$, and hence $d_1 \leq g + d_2$. Thus,

$$\deg_u(abh) \leq d_1 + d_2 + g \leq 2g + 2d_2 < 2g + 2d_2 + 1 = \deg_u(b^2 f) \ .$$

It follows that

$$\deg_u(N(G)) = \max(2d_1, 2g + 1 + 2d_2) = \deg(G) \ .$$

2) We have

$$\begin{aligned}
\deg(GH) &= \deg_u(N(GH)) \ , \quad \text{by 1)} \\
&= \deg_u(N(G)N(H)) \ , \quad \text{by part 3) of Lemma 2.2} \\
&= \deg_u(N(G)) + \deg_u(N(H)) \\
&= \deg(G) + \deg(H) \ .
\end{aligned}$$

3) Since $N(G) = N(\overline{G})$, we have $\deg(G) = \deg_u(N(G)) = \deg_u(N(\overline{G})) = \deg(\overline{G})$. □

Definition 2.7. Let $R = G/H \in \overline{\mathbb{F}}(C)$ be a rational function.

1) If $\deg(G) < \deg(H)$ then the value of R at ∞ is defined to be $R(\infty) = 0$.

2) If $\deg(G) > \deg(H)$ then R is *not defined* at ∞.

3) If $\deg(G) = \deg(H)$ then $R(\infty)$ is defined to be the ratio of the leading coefficients (with respect to the deg function) of G and H.

§ 3. Zeros and Poles

This section introduces the notion of a uniformizing parameter, and the orders of zeros and poles of rational functions.

Definition 3.1. Let $R \in \overline{\mathbb{F}}(C)$ be a nonzero rational function, and let $P \in C$. If $R(P) = 0$ then R is said to have a *zero* at P. If R is not defined at P then R is said to have a *pole* at P, in which case we write $R(P) = \infty$.

Lemma 3.1. *Let $G \in \overline{\mathbb{F}}[C]$ be a nonzero polynomial function, and let $P \in C$. If $G(P) = 0$, then $\overline{G}(\tilde{P}) = 0$.*

Proof. Let $G = a(u) - b(u)v$ and $P = (x, y)$. Then $\overline{G} = a(u) + b(u)(v + h(u))$, $\widetilde{P} = (x, -y - h(x))$, and $\overline{G}(\widetilde{P}) = a(x) + b(x)(-y - h(x) + h(x)) = a(x) - yb(x) = G(P) = 0$. $\square$

The next three lemmas are used in the proof of Theorem 3.1, which establishes the existence of uniformizing parameters.

Lemma 3.2. *Let $P = (x, y)$ be a point on C. Suppose that a nonzero polynomial function $G = a(u) - b(u)v \in \overline{\mathbb{F}}[C]$ has a zero at P, and suppose that x is not a root of both $a(u)$ and $b(u)$. Then $\overline{G}(P) = 0$ if and only if P is a special point.*

Proof. If P is a special point, then $\overline{G}(P) = 0$ by Lemma 3.1. Conversely, suppose that P is an ordinary point, i.e., $y \neq (-y - h(x))$. If $\overline{G}(P) = 0$ then we have:

$$a(x) - b(x)y = 0$$
$$a(x) + b(x)(h(x) + y) = 0 \ .$$

Subtracting the two equations, we obtain $b(x) = 0$, and hence $a(x) = 0$, which contradicts the hypothesis that x is not a root of both $a(u)$ and $b(u)$. Hence if $\overline{G}(P) = 0$, it follows that P is special. $\square$

Lemma 3.3. *Let $P = (x, y)$ be an ordinary point on C, and let $G = a(u) - b(u)v \in \overline{\mathbb{F}}[C]$ be a nonzero polynomial function. Suppose that $G(P) = 0$ and x is not a root of both $a(u)$ and $b(u)$. Then G can be written in the form $(u - x)^s S$, where s is the highest power of $(u - x)$ that divides $N(G)$, and $S \in \overline{\mathbb{F}}(C)$ has neither a zero nor a pole at P.*

Proof. We can write

$$G = G \cdot \frac{\overline{G}}{\overline{G}} = \frac{N(G)}{\overline{G}} = \frac{a^2 + abh - b^2 f}{a + b(h + v)} \ .$$

Let $N(G) = (u - x)^s d(u)$, where s is the highest power of $(u - x)$ that divides $N(G)$ (so $d(u) \in \overline{\mathbb{F}}[u]$ and $d(x) \neq 0$). By Lemma 3.2, $\overline{G}(P) \neq 0$. Let $S = d(u)/\overline{G}$. Then $G = (u - x)^s S$ and $S(P) \neq 0, \infty$. $\square$

Lemma 3.4. *Let $P = (x, y)$ be a special point on C. Then $(u - x)$ can be written in the form $(v - y)^2 \cdot S(u, v)$, where $S(u, v) \in \overline{\mathbb{F}}(C)$ has neither a zero nor a pole at P.*

Proof. Let $H = (v - y)^2$ and $S = (u - x)/H$, so that $(u - x) = H \cdot S$. We will show that $S(P) \neq 0, \infty$. Since P is a special point, $2y + h(x) = 0$. Consequently, since P is not a singular point, we have $h'(x)y - f'(x) \neq 0$. Also, $f(x) = y^2 + h(x)y = y^2 + (-2y)(y) = -y^2$. Now,

$$H(u, v) = (v - y)^2 = v^2 - 2yv + y^2 = f(u) - h(u)v - 2yv + y^2 \ .$$

Hence

$$\frac{1}{S(u, v)} = \left(\frac{f(u) + y^2}{u - x} \right) - v \left(\frac{h(u) + 2y}{u - x} \right) \ . \tag{2}$$

Notice that the right hand side of (2) is indeed a polynomial function. Let $s(u) = H(u, y)$, and observe that $s(x) = 0$. Moreover, $s'(u) = f'(u) - h'(u)y$, whence $s'(x) \neq 0$. Thus $(u - x)$ divides $s(u)$, but $(u - x)^2$ does not divide $s(u)$. It follows that the right hand side of (2) is nonzero at P, and hence that $S(P) \neq 0, \infty$, as required. $\square$

Theorem 3.1. *Let $P \in C$. Then there exists a function $U \in \overline{\mathbb{F}}(C)$ with $U(P) = 0$ such that the following property holds: for each nonzero polynomial function $G \in \overline{\mathbb{F}}[C]$, there exist an integer d and a function $S \in \overline{\mathbb{F}}(C)$ such that $S(P) \neq 0, \infty$ and $G = U^d S$. Furthermore, the number d does not depend on the choice of U. The function U is called a uniformizing parameter for P.*

Proof. Let $G(u, v) \in \overline{\mathbb{F}}[C]$ be a nonzero polynomial function. If P is a finite point, suppose that $G(P) = 0$; if $P = \infty$, suppose that $G(P) = \infty$. (If $G(P) \neq 0, \infty$, then we can write $G = U^0 G$ where U is any polynomial in $\overline{\mathbb{F}}[C]$ satisfying $U(P) = 0$.) We prove the theorem by finding a uniformizing parameter for each of the following cases: 1) $P = \infty$; 2) P is an ordinary point; and 3) P is a special point.

1) We show that a uniformizing parameter for the point $P = \infty$ is $U = u^g/v$. First note that $U(\infty) = 0$ since $\deg(u^g) < \deg(v)$. Next, write

$$G = \left(\frac{u^g}{v}\right)^d \left(\frac{v}{u^g}\right)^d G ,$$

where $d = -\deg(G)$. Let $S = (v/u^g)^d G$. Since $\deg(v) - \deg(u^g) = 2g + 1 - 2g = 1$ and $d = -\deg(G)$, it follows that $\deg(u^{-gd}G) = \deg(v^{-d})$. Hence $S(\infty) \neq 0, \infty$.

2) Assume now that $P = (x, y)$ is an ordinary point. We show that a uniformizing parameter for P is $U = (u - x)$; observe that $U(P) = 0$. Write $G = a(u) - b(u)v$. Let $(u - x)^r$ be the highest power of $(u - x)$ which divides both $a(u)$ and $b(u)$, and write

$$G(u, v) = (u - x)^r (a_0(u) - b_0(u)v) .$$

By Lemma 3.3, we can write $(a_0(u) - b_0(u)v) = (u - x)^s S$ for some integer $s \geq 0$ and some $S \in \overline{\mathbb{F}}(C)$ such that $S(P) \neq 0, \infty$. Hence $G = (u - x)^{r+s} S$ satisfies the conclusion of the theorem with $d = r + s$.

3) Assume now that $P = (x, y)$ is a special point. We show that a uniformizing parameter for P is $U = (v - y)$; observe that $U(P) = 0$. By replacing any powers of u greater than $2g$ with the equation of the curve, we can write

$$G(u, v) = u^{2g}b_{2g}(v) + u^{2g-1}b_{2g-1}(v) + \cdots + ub_1(v) + b_0(v) ,$$

where each $b_i(v) \in \overline{\mathbb{F}}[v]$. Replacing all occurrences of u by $((u - x) + x)$ and expanding, we obtain

$$G(u, v) = (u - x)^{2g}\overline{b}_{2g}(v) + (u - x)^{2g-1}\overline{b}_{2g-1}(v) + \cdots + (u - x)\overline{b}_1(v) + \overline{b}_0(v)$$

$$= (u - x)B(u, v) + \overline{b}_0(v) ,$$

where each $\bar{b}_i(v) \in \bar{\mathbb{F}}[v]$, and $B(u, v) \in \bar{\mathbb{F}}[C]$. Now $G(P) = 0$ implies $\bar{b}_0(y) = 0$, and so we can write $\bar{b}_0(v) = (v - y)c(v)$ for some $c \in \bar{\mathbb{F}}[v]$. By the proof of Lemma 3.4 (see equation (2)), we can write $(u - x) = (v - y)^2 / A(u, v)$, where $A(u, v) \in \bar{\mathbb{F}}[C]$ and $A(P) \neq 0, \infty$. Hence

$$G(u, v) = (v - y)\left[\frac{(v - y)B(u, v)}{A(u, v)} + c(v)\right]$$

$$= \frac{(v - y)}{A(u, v)}[(v - y)B(u, v) + A(u, v)c(v)]$$

$$\overset{\text{def}}{=} \frac{(v - y)}{A(u, v)}G_1(u, v) \ .$$

Now if $G_1(P) \neq 0$, then we are done, since we can take $S = G_1/A$. On the other hand, if $G_1(P) = 0$, then $c(y) = 0$ and we can write $c(v) = (v - y)c_1(v)$ for some $c_1 \in \bar{\mathbb{F}}[v]$. Hence

$$G = (v - y)^2\left[\frac{B(u, v)}{A(u, v)} + c_1(v)\right]$$

$$= \frac{(v - y)^2}{A(u, v)}\left[B(u, v) + A(u, v)c_1(v)\right]$$

$$\overset{\text{def}}{=} \frac{(v - y)^2}{A(u, v)}G_2(u, v) \ .$$

Again, if $G_2(P) \neq 0$, then we are done. Otherwise, the whole process can be repeated. To see that the process terminates, suppose that we have pulled out k factors of $v - y$. There are two cases to consider.

a) If k is even, say $k = 2l$, we can write

$$G = \frac{(v - y)^{2l}}{A(u, v)^l} D(u, v)$$

where $D \in \bar{\mathbb{F}}[C]$. Hence, $A^l G = (v - y)^{2l} D = (u - x)^l A^l D$, whence $G = (u-x)^l D$. Taking norms of both sides, we have $N(G) = (u-x)^{2l} N(D)$. Hence $k \leq \deg_u(N(G))$.

b) If k is odd, say $k = 2l + 1$, we can write

$$G = \frac{(v - y)^{2l+1}}{A(u, v)^{l+1}} D(u, v) \ ,$$

where $D \in \bar{\mathbb{F}}[C]$. Hence, $A^{l+1} G = (v - y)^{2l+1} D = (u - x)^l A^l(v - y)D$, whence $AG = (u - x)^l(v - y)D$. Taking norms of both sides, we have $N(AG) = (u - x)^{2l} N(v - y)N(D)$. Hence $2l < \deg_u(N(AG))$, and so $k \leq \deg_u(N(AG))$.

In either case, k is bounded by $\deg_u(N(AG))$, and so the process must terminate.

To see that d is independent of the choice of U, suppose that U_1 is another uniformizing parameter for P. Since $U(P) = U_1(P) = 0$, we can write $U = U_1^a A$

and $U_1 = U^b B$, where $a \geq 1$, $b \geq 1$, $A, B \in \overline{\mathbb{F}}(C)$, $A(P) \neq 0, \infty$, $B(P) \neq 0, \infty$. Thus $U = (U^b B)^a A = U^{ab} B^a A$. Dividing both sides by U, we obtain $U^{ab-1} B^a A = 1$. If we substitute P in both sides of this equation, we see that $ab - 1 = 0$. Hence $a = b = 1$. Thus $G = U^d S = U_1^d (A^d S)$, where $A^d S$ has neither a zero nor a pole at P. $\square$

The notion of a uniformizing parameter is next used to define the order of a polynomial function at a point. An alternative definition from [Koblitz 1989], which is more convenient to use for computational purposes, is given in Definition 3.3. Lemma 3.6 establishes that these two definitions are in fact equivalent.

Definition 3.2. Let $G \in \overline{\mathbb{F}}[C]$ be a nonzero polynomial function, and let $P \in C$. Let $U \in \overline{\mathbb{F}}(C)$ be a uniformizing parameter for P, and write $G = U^d S$ where $S \in \overline{\mathbb{F}}(C)$, $S(P) \neq 0, \infty$. The *order of G at P* is defined to be $\mathrm{ord}_P(G) = d$.

Lemma 3.5. *Let $G_1, G_2 \in \overline{\mathbb{F}}[C]$ be nonzero polynomial functions, and let $P \in C$. Let $\mathrm{ord}_P(G_1) = r_1$, $\mathrm{ord}_P(G_2) = r_2$.*
1) $\mathrm{ord}_P(G_1 G_2) = \mathrm{ord}_P(G_1) + \mathrm{ord}_P(G_2)$.
2) *If $r_1 \neq r_2$, then $\mathrm{ord}_P(G_1 + G_2) = \min(r_1, r_2)$. If $r_1 = r_2$ and $G_1 \neq -G_2$, then $\mathrm{ord}_P(G_1 + G_2) \geq r_2$.*

Proof. Let U be a uniformizing parameter for P. By Definition 3.2, we can write $G_1 = U^{r_1} S_1$ and $G_2 = U^{r_2} S_2$, where $S_1, S_2 \in \overline{\mathbb{F}}(C)$, $S_1(P) \neq 0, \infty$, $S_2(P) \neq 0, \infty$. Without loss of generality, suppose that $r_1 \geq r_2$.
1) $G_1 G_2 = U^{r_1 + r_2}(S_1 S_2)$, from which it follows that $\mathrm{ord}_P(G_1 G_2) = r_1 + r_2$.
2) $G_1 + G_2 = U^{r_2}(U^{r_1 - r_2} S_1 + S_2)$. If $r_1 > r_2$, then $(U^{r_1 - r_2} S_1)(P) = 0$, $S_2(P) \neq 0, \infty$, and so $\mathrm{ord}_P(G_1 + G_2) = r_2$. If $r_1 = r_2$, then $(S_1 + S_2)(P) \neq \infty$ (although it may be the case that $(S_1 + S_2)(P) = 0$), and so $\mathrm{ord}_P(G_1 + G_2) \geq r_2$. $\square$

We now give an alternate definition of the order of a polynomial function at a point.

Definition 3.3. Let $G = a(u) - b(u)v \in \overline{\mathbb{F}}[C]$ be a nonzero polynomial function, and let $P \in C$. The *order of G at P*, denoted $\mathrm{ord}_P(G)$, is defined as follows:
1) If $P = (x, y)$ is a finite point, then let r be the highest power of $(u - x)$ that divides both $a(u)$ and $b(u)$, and write $G(u, v) = (u - x)^r(a_0(u) - b_0(u)v)$. If $a_0(x) - b_0(x)y \neq 0$, then let $s = 0$; otherwise, let s be the highest power of $(u - x)$ that divides $N(a_0(u) - b_0(u)v) = a_0^2 + a_0 b_0 h - b_0^2 f$. If P is an ordinary point, then define $\mathrm{ord}_P(G) = r + s$. If P is a special point, then define $\mathrm{ord}_P(G) = 2r + s$.
2) If $P = \infty$, then

$$\mathrm{ord}_P(G) = -\max\{2\deg_u(a),\ 2g + 1 + 2\deg_u(b)\}\ .$$

Lemma 3.6. *Definitions 3.2 and 3.3 are equivalent. That is, if the order function of Definition 3.3 is denoted by $\overline{\mathrm{ord}}$, then $\mathrm{ord}_P(G) = \overline{\mathrm{ord}}_P(G)$ for all $P \in C$ and all nonzero $G \in \overline{\mathbb{F}}[C]$.*

Proof. If $P = \infty$, the lemma follows directly from the proof of part 1) of Theorem 3.1. For the case when P is an ordinary point, the lemma follows directly from Lemma 3.3 and the proof of part 2) of Theorem 3.1.

Suppose now that $P = (x, y)$ is a special point, and let $G = a - bv$. Let r be the highest power of $(u - x)$ which divides both $a(u)$ and $b(u)$, and write

$$G = (u - x)^r (a_0(u) - b_0(u)v) \stackrel{\text{def}}{=} (u - x)^r H(u, v) .$$

Let $\operatorname{ord}_P(H) = s$. Then, by Lemma 3.4,

$$\operatorname{ord}_P(G) = \operatorname{ord}_P((u - x)^r) + \operatorname{ord}_P(H) = 2r + s .$$

Now since $v - y$ is a uniformizing parameter for P, we can write

$$H(u, v) = (v - y)^s A_1/A_2 , \quad \text{where } A_1, A_2 \in \mathbb{F}[C] , \quad A_1(P) \neq 0 , \quad A_2(P) \neq 0 .$$

Multiplying both sides by A_2 and taking norms, we have

$$N(A_2)N(H) = (y^2 + h(u)y - f(u))^s N(A_1) .$$

Now $N(A_1)(x) \neq 0$, since $A_1(P) \neq 0$ and P is special (Lemma 3.1). Similarly, $N(A_2)(x) \neq 0$. Also, $u = x$ is a root of the polynomial $y^2 + h(u)y - f(u)$. Moreover, $u = x$ is not a double root of $y^2 + h(u)y - f(u)$, since $h'(x)y - f'(x) \neq 0$. It follows that $(u - x)^s$ is the highest power of $(u - x)$ that divides $N(H)$. Hence, $\overline{\operatorname{ord}}_P(G) = 2r + s = \operatorname{ord}_P(G)$. $\square$

Lemma 3.7 is a generalization of Lemma 3.1.

Lemma 3.7. *Let $G \in \mathbb{F}[C]$ be a nonzero polynomial function, and let $P \in C$. Then $\operatorname{ord}_P(G) = \operatorname{ord}_{\tilde{P}}(\overline{G})$.*

Proof. There are two cases to consider.
1) Suppose $P = \infty$; then $\tilde{P} = \infty$. By Definition 2.6 and part 2) of Definition 3.3, $\operatorname{ord}_P(G) = -\deg(G)$ and $\operatorname{ord}_{\tilde{P}}(\overline{G}) = \operatorname{ord}_P(\overline{G}) = -\deg(\overline{G})$. By part 3) of Lemma 2.3, $\deg(G) = \deg(\overline{G})$. Hence, $\operatorname{ord}_P(G) = \operatorname{ord}_{\tilde{P}}(\overline{G})$.
2) Suppose now that $P = (x, y)$ is a finite point. Let $G = a(u) - b(u)v = (u - x)^r H(u, v)$, where r is the highest power of $(u - x)$ that divides both $a(u)$ and $b(u)$ and $H(u, v) = a_0(u) - b_0(u)v$. If $H(x, y) \neq 0$, then let $s = 0$; otherwise, let s be the highest power of $(u - x)$ that divides $N(H)$. Now $\overline{G} = (u - x)^r \overline{H}$, where $\overline{H} = (a_0 + b_0 h) + b_0 v$. Recall that $H(P) = 0$ if and only if $\overline{H}(\tilde{P}) = 0$. Since $(u - x)$ does not divide both $a_0 + b_0 h$ and b_0 (since otherwise, $(u - x)|a_0$), and s is the highest power of $(u - x)$ that divides $N(H) = N(\overline{H})$, it follows from Definition 3.3 that $\operatorname{ord}_{\tilde{P}}(\overline{G}) = \operatorname{ord}_P(G)$. $\square$

Theorem 3.2. *Let $G \in \mathbb{F}[C]$ be a nonzero polynomial function. Then G has a finite number of zeros and poles. Moreover, $\sum_{P \in C} \operatorname{ord}_P(G) = 0$.*

Proof. Let $n = \deg(G)$; then $\deg_u(N(G)) = n$. We can write

$$N(G) = G\overline{G} = (u - x_1)(u - x_2) \cdots (u - x_n) ,$$

where $x_i \in \overline{\mathbb{F}}$, and the x_i are not necessarily distinct. The only pole of G is at $P = \infty$, and $\text{ord}_\infty(G) = -n$. If x_i is the u-coordinate of an ordinary point $P = (x_i, y_i)$ on C, then $\text{ord}_P(u - x_i) = 1$ and $\text{ord}_{\tilde{P}}(u - x_i) = 1$, and $(u - x_i)$ has no other zeros. If x_i is the u-coordinate of a special point $P = (x_i, y_i)$ on C, then $\text{ord}_P(u - x_i) = 2$, and $(u - x_i)$ has no other zeros. Hence, $N(G)$, and consequently also G, has a finite number of zeros and poles, and moreover $\sum_{P \in C \setminus \{\infty\}} \text{ord}_P(N(G)) = 2n$. But, by Lemma 3.7, $\sum_{P \in C \setminus \{\infty\}} \text{ord}_P(G) = \sum_{P \in C \setminus \{\infty\}} \text{ord}_P(\overline{G})$, and hence $\sum_{P \in C \setminus \{\infty\}} \text{ord}_P(G) = n$. We conclude that $\sum_{P \in C} \text{ord}_P(G) = 0$. $\square$

Definition 3.4. Let $R = G/H \in \overline{\mathbb{F}}(C)$ be a nonzero rational function, and let $P \in C$. The *order of R at P* is defined to be $\text{ord}_P(R) = \text{ord}_P(G) - \text{ord}_P(H)$.

It can readily be verified that $\text{ord}_P(R)$ does not depend on the choice of G and H, and that Lemma 3.5 and Theorem 3.2 are also true for nonzero rational functions.

§ 4. Divisors

This section presents the basic properties of divisors and introduces the jacobian of a hyperelliptic curve.

Definition 4.1. A *divisor* D is a formal sum of points on C

$$D = \sum_{P \in C} m_P P , \quad m_P \in \mathbb{Z} ,$$

where only a finite number of the integers m_P are nonzero. The *degree* of D, denoted $\deg D$, is the integer $\sum_{P \in C} m_P$. The *order* of D at P is the integer m_P; we write $\text{ord}_P(D) = m_P$.

The set of all divisors, denoted $\mathbb{D}$, forms an additive group under the addition rule:

$$\sum_{P \in C} m_P P + \sum_{P \in C} n_P P = \sum_{P \in C} (m_P + n_P)P .$$

The set of all divisors of degree 0, denoted $\mathbb{D}^0$, is a subgroup of $\mathbb{D}$.

Definition 4.2. Let $D_1 = \sum_{P \in C} m_P P$ and $D_2 = \sum_{P \in C} n_P P$ be two divisors. The *greatest common divisor* of D_1 and D_2 is defined to be

$$\text{g.c.d.}(D_1, D_2) = \sum_{P \in C} \min(m_P, n_P)P - \left(\sum_{P \in C} \min(m_P, n_P) \right) \infty .$$

(Note that $\text{g.c.d.}(D_1, D_2) \in \mathbb{D}^0$.)

Definition 4.3. Let $R \in \overline{\mathbb{F}}(C)$ be a nonzero rational function. The *divisor of R* is

$$\text{div}(R) = \sum_{P \in C} (\text{ord}_P R)P .$$

Note that if $R = G/H$ then $\text{div}(R) = \text{div}(G) - \text{div}(H)$. Theorem 3.2 shows that the divisor of a rational function is indeed a finite formal sum and has degree 0.

Example 4.1. If $P = (x, y)$ is an ordinary point on C, then $\text{div}(u-x) = P + \widetilde{P} - 2\infty$. If $P = (x, y)$ is a special point on C, then $\text{div}(u - x) = 2P - 2\infty$.

Lemma 4.1. *Let $G \in \overline{\mathbb{F}}[C]$ be a nonzero polynomial function, and let $\text{div}(G) = \sum_{P \in C} m_P P$. Then $\text{div}(\overline{G}) = \sum_{P \in C} m_P \widetilde{P}$.*

Proof. The result follows directly from Lemma 3.7. $\square$

If $R_1, R_2 \in \overline{\mathbb{F}}(C)$ are nonzero rational functions, then it follows from part 1) of Lemma 3.5 that $\text{div}(R_1 R_2) = \text{div}(R_1) + \text{div}(R_2)$.

Definition 4.4. A divisor $D \in \mathbb{D}^0$ is called a *principal divisor* if $D = \text{div}(R)$ for some nonzero rational function $R \in \overline{\mathbb{F}}(C)$. The set of all principal divisors, denoted $\mathbb{P}$, is a subgroup of $\mathbb{D}^0$. The quotient group $\mathbb{J} = \mathbb{D}^0/\mathbb{P}$ is called the *jacobian* of the curve C. If $D_1, D_2 \in \mathbb{D}^0$ then we write $D_1 \sim D_2$ if $D_1 - D_2 \in \mathbb{P}$; D_1 and D_2 are said to be *equivalent* divisors.

Definition 4.5. Let $D = \sum_{P \in C} m_P P$ be a divisor. The *support* of D is the set $\text{supp}(D) = \{P \in C \mid m_P \neq 0\}$.

Definition 4.6. A *semi-reduced divisor* is a divisor of the form $D = \sum m_i P_i - (\sum m_i)\infty$, where each $m_i \geq 0$ and the P_i's are finite points such that when $P_i \in \text{supp}(D)$ one has $\widetilde{P}_i \notin \text{supp}(D)$, unless $P_i = \widetilde{P}_i$, in which case $m_i = 1$.

Lemma 4.2. *For each divisor $D \in \mathbb{D}^0$ there exists a semi-reduced divisor $D_1 \in \mathbb{D}^0$ such that $D \sim D_1$.*

Proof. Let $D = \sum_{P \in C} m_P P$. Let (C_1, C_2) be a partition of the set of ordinary points on C such that 1) $P \in C_1$ if and only if $\widetilde{P} \in C_2$; and 2) if $P \in C_1$ then $m_P \geq m_{\widetilde{P}}$. Let C_0 be the set of special points on C. Then we can write

$$D = \sum_{P \in C_1} m_P P + \sum_{P \in C_2} m_P P + \sum_{P \in C_0} m_P P - m\infty .$$

Consider the following divisor

$$D_1 = D - \sum_{P=(x,y) \in C_2} m_P \text{div}(u - x) - \sum_{P=(x,y) \in C_0} \left[\frac{m_P}{2}\right] \text{div}(u - x) .$$

Then $D_1 \sim D$. Finally, by Example 4.1, we have

$$D_1 = \sum_{P \in C_1} (m_P - m_{\widetilde{P}})P + \sum_{P \in C_0} \left(m_P - 2\left[\frac{m_P}{2}\right]\right) P - m_1\infty$$

for some integer $m_1 \geq 0$, and hence D_1 is a semi-reduced divisor. $\square$

§5. Representing Semi-Reduced Divisors

This section describes a polynomial representation for semi-reduced divisors of the jacobian. It leads to an efficient algorithm for adding elements of the jacobian (see §7).

Lemma 5.1. *Let* $P = (x, y)$ *be an ordinary point on* C, *and let* $R \in \overline{\mathbb{F}}(C)$ *be a rational function that does not have a pole at* P. *Then for any* $k \geq 0$, *there are unique elements* $c_0, c_1, \ldots, c_k \in \overline{\mathbb{F}}$ *and* $R_k \in \overline{\mathbb{F}}(C)$ *such that* $R = \sum_{i=0}^{k} c_i(u - x)^i + (u - x)^{k+1} R_k$, *where* R_k *does not have a pole at* P.

Proof. There is a unique $c_0 \in \overline{\mathbb{F}}$, namely $c_0 = R(x, y)$, such that P is a zero of $R - c_0$. Since $(u - x)$ is a uniformizing parameter for P, we can write $R - c_0 = (u - x)R_1$ for some (unique) $R_1 \in \overline{\mathbb{F}}(C)$ with $\mathrm{ord}_P(R_1) \geq 0$. Hence $R = c_0 + (u - x)R_1$. The lemma now follows by induction. $\square$

In the next lemma, when we write "mod $(u - x)^k$", we mean modulo the ideal generated by $(u - x)^k$ in the subring of $\overline{\mathbb{F}}(C)$ consisting of rational functions that do not have a pole at P. Thus, the conclusion in Lemma 5.1 can be restated: $R \equiv \sum_{i=0}^{k} c_i(u - x)^k \pmod{(u - x)^{k+1}}$.

Lemma 5.2. *Let* $P = (x, y)$ *be an ordinary point on* C. *Then for each* $k \geq 1$, *there exists a unique polynomial* $b_k(u) \in \overline{\mathbb{F}}[u]$ *such that*
1) $\deg_u b_k < k$;
2) $b_k(x) = y$; *and*
3) $b_k^2(u) + b_k(u)h(u) \equiv f(u) \pmod{(u - x)^k}$.

Proof. We apply Lemma 5.1 to $R(u, v) = v$. Let $v = \sum_{i=0}^{k-1} c_i(u - x)^i + (u - x)^k R_{k-1}$, where $c_i \in \overline{\mathbb{F}}$ and $R_{k-1} \in \overline{\mathbb{F}}(C)$. Define $b_k(u) = \sum_{i=0}^{k-1} c_i(u - x)^i$. From the proof of Lemma 5.1, we know that $c_0 = y$, and hence $b_k(x) = y$. Finally, since $v^2 + h(u)v = f(u)$, if we reduce both sides modulo $(u - x)^k$ we obtain $b_k(u)^2 + b_k(u)h(u) \equiv f(u) \pmod{(u - x)^k}$. Uniqueness is easily proved by induction on k. $\square$

The following theorem shows how a semi-reduced divisor can be represented as the g.c.d. of the divisors of two polynomial functions.

Theorem 5.1. *Let* $D = \sum m_i P_i - (\sum m_i)\infty$ *be a semi-reduced divisor, where* $P_i = (x_i, y_i)$. *Let* $a(u) = \prod(u - x_i)^{m_i}$. *There exists a unique polynomial* $b(u)$ *satisfying:* 1) $\deg_u b < \deg_u a$; 2) $b(x_i) = y_i$ *for all* i *for which* $m_i \neq 0$; *and* 3) $a(u)$ *divides* $(b(u)^2 + b(u)h(u) - f(u))$. *Then* $D = \mathrm{g.c.d.}(\mathrm{div}(a(u)), \mathrm{div}(b(u) - v))$.

Notation: $\mathrm{g.c.d.}(\mathrm{div}(a(u)), \mathrm{div}(b(u) - v))$ will usually be abbreviated to $\mathrm{div}(a(u), b(u) - v)$ or, more simply, to $\mathrm{div}(a, b)$.

Proof. Let C_1 be the set of ordinary points in $\mathrm{supp}(D)$, and let C_0 be the set of special points in $\mathrm{supp}(D)$. Let $C_2 = \{\tilde{P} : P \in C_1\}$. Then we can write

$$D = \sum_{P_i \in C_0} P_i + \sum_{P_i \in C_1} m_i P_i - m\infty \ ,$$

where m_i, m are positive integers.

We first prove that there exists a unique polynomial $b(u)$ which satisfies the conditions of the theorem. By Lemma 5.2, for each $P_i \in C_1$ there exists a unique polynomial $b_i(u) \in \overline{\mathbb{F}}[u]$ satisfying 1) $\deg_u b_i < m_i$; 2) $b_i(x_i) = y_i$; and 3) $(u - x_i)^{m_i} | b_i^2(u) + b_i(u)h(u) - f(u)$. It can easily be verified that for each $P_i \in C_0$, $b_i(u) = y_i$ is the unique polynomial satisfying 1) $\deg_u b_i < 1$; 2) $b_i(x_i) = y_i$; and 3) $(u - x_i) | b_i^2(u) + b_i(u)h(u) - f(u)$. By the Chinese Remainder Theorem for polynomials (see Exercise 3 in §3 of Chapter 3), there is a unique polynomial $b(u) \in \overline{\mathbb{F}}[u]$, $\deg_u b < \sum m_i$, such that

$$b(u) \equiv b_i(u) \ (\mathrm{mod}(u - x_i)^{m_i}) \ \text{for all } i \ .$$

It can now be verified that $b(u)$ satisfies conditions 1), 2) and 3) of the theorem.

Next,

$$\mathrm{div}(a(u)) = \mathrm{div}\left(\prod (u - x_i)^{m_i}\right) = \sum_{P_i \in C_0} 2P_i + \sum_{P_i \in C_1} m_i P_i + \sum_{P_i \in C_1} m_i \tilde{P}_i - (*)\infty \ .$$

In addition,

$$\mathrm{div}(b(u) - v) = \sum_{P_i \in C_0} t_i P_i + \sum_{P_i \in C_1} s_i P_i + \sum_{P_i \in C \backslash (C_0 \cup C_1 \cup C_2 \cup \{\infty\})} m_i P_i - (*)\infty \ ,$$

where each $s_i \geq m_i$ since $(u - x_i)^{m_i}$ divides $N(b - v) = b^2 + hb - f$. Now if $P = (x, y) \in C_0$, then $(u - x)$ divides $b^2 + bh - f$. The derivative of this polynomial evaluated at $u = x$ is

$$2b(x)b'(x) + b'(x)h(x) + b(x)h'(x) - f'(x)$$
$$= b'(x)(2y + h(x)) + (h'(x)y - f'(x))$$
$$= h'(x)y - f'(x) \ , \quad \text{since } 2y + h(x) = 0$$
$$\neq 0 \ .$$

Thus, $u = x$ is a simple root of $N(b - v) = b^2 + bh - f$, and hence $t_i = 1$ for all i. Therefore,

$$\mathrm{g.c.d.}(a(u), b(u) - v) = \sum_{P_i \in C_0} P_i + \sum_{P_i \in C_1} m_i P_i - m\infty = D \ ,$$

as required. □

Note that the zero divisor is represented as $\mathrm{div}(1, 0)$. The next result follows from the proof of Theorem 5.1.

Lemma 5.3. *Let $a(u), b(u) \in \overline{\mathbb{F}}[u]$ be such that $\deg_u b < \deg_u a$. If $a | (b^2 + bh - f)$, then $\mathrm{div}(a, b)$ is semi-reduced.*

§ 6. Reduced Divisors

This section defines the notion of a reduced divisor and proves that each coset in the quotient group $\mathbb{J} = \mathbb{D}^0/\mathbb{P}$ has exactly one reduced divisor. We can therefore identify each element of $\mathbb{J}$ with its reduced divisor.

Definition 6.1. Let $D = \sum m_i P_i - (\sum m_i)\infty$ be a semi-reduced divisor. If $\sum m_i \le g$ (g is the genus of C) then D is called a *reduced divisor*.

Definition 6.2. Let $D = \sum_{P \in C} m_P P$ be a divisor. The *norm* of D is defined to be

$$|D| = \sum_{P \in C \backslash \{\infty\}} |m_P| \; .$$

Note that given a divisor $D \in \mathbb{D}^0$, the operation described in the proof of Lemma 4.2 produces a semi-reduced divisor D_1 such that $D_1 \sim D$ and $|D_1| \le |D|$.

Lemma 6.1. *Let R be a nonzero rational function in $\overline{\mathbb{F}}(C)$. If R has no finite poles, then R is a polynomial function.*

Proof. Let $R = G/H$, where G, H are nonzero polynomial functions in $\overline{\mathbb{F}}[C]$. Then $R = \frac{G}{H} \cdot \frac{\overline{H}}{\overline{H}} = G\overline{H}/N(H)$, and so we can write $R = (a - bv)/c$, where $a, b, c \in \overline{\mathbb{F}}[u]$, $c \ne 0$. Let $x \in \overline{\mathbb{F}}$ be a root of c. Let $P = (x, y) \in C$ where $y \in \overline{\mathbb{F}}$, and let $d \ge 1$ be the highest power of $(u - x)$ that divides c.

If P is ordinary, then $\text{ord}_P(c) = \text{ord}_{\widetilde{P}}(c) = d$. Since R has no finite poles, $\text{ord}_P(a - bv) \ge d$ and $\text{ord}_{\widetilde{P}}(a - bv) \ge d$. Now since P and $\widetilde{P}$ are both zeros of $a - bv$, we have $a(x) = 0$ and $b(x) = 0$. It follows that $\text{ord}_P(a) \ge d$ and $\text{ord}_P(b) \ge d$. Hence $(u - x)^d$ is a common divisor of a and b, and it can be canceled with the factor $(u - x)^d$ of c.

Suppose now that P is special. Then $\text{ord}_P(c) = 2d$. Since R has no finite poles, $\text{ord}_P(a - bv) \ge 2d$. Then, as in part 3) of the proof of Theorem 3.1, we can write

$$a - bv = \frac{(v - y)^{2d} D}{A^d} \; ,$$

where A and D are nonzero polynomial functions in $\overline{\mathbb{F}}[C]$, and A satisfies $(v-y)^2 = (u - x)A$. Hence $a - bv = (u - x)^d D$. Again, the factor $(u - x)^d$ of $a - bv$ can be canceled with the factor $(u - x)^d$ of c.

This can be repeated for all roots of c; it follows that R is a polynomial function. $\square$

Theorem 6.1. *For each divisor $D \in \mathbb{D}^0$ there exists a unique reduced divisor D_1 such that $D \sim D_1$.*

Proof. Existence. Let D' be a semi-reduced divisor such that $D' \sim D$ and $|D'| \le |D|$ (see the proof of Lemma 4.2). If $|D'| \le g$, then D' is reduced and we are done. Otherwise, let $P_1, P_2, \ldots, P_{g+1}$ be finite points in $\text{supp}(D')$. The points P_i are not necessarily distinct, but a point P cannot occur in this list more than $\mathrm{ord}_P(D')$ times. Let $\mathrm{div}(a(u), b(u))$ be the representation of the divisor

$$P_1 + P_2 + \cdots + P_{g+1} - (g+1)\infty$$

given by Theorem 5.1. Since $\deg_u(b) \leq g$, we have $\deg(b(u) - v) = 2g + 1$, and hence

$$\mathrm{div}(b(u) - v) = P_1 + P_2 + \cdots + P_{g+1} + Q_1 + \cdots + Q_g - (2g+1)\infty$$

for some finite points $Q_1, Q_2, \ldots, Q_g$. Subtracting this divisor from D' gives a divisor D'', where $D'' \sim D' \sim D$ and $|D''| < |D'|$. We can now produce another semi-reduced divisor $D''' \sim D''$ such that $|D'''| \leq |D''|$. After doing this a finite number of times, we obtain a semi-reduced divisor D_1 with $|D_1| \leq g$, and we are done.

Algorithm 2 in §7 describes an efficient algorithm which, given a semi-reduced divisor $D = \mathrm{div}(a, b)$, finds a reduced divisor D_1 such that $D \sim D_1$; the algorithm only uses a and b.

Uniqueness. Suppose that D_1 and D_2 are two reduced divisors with $D_1 \sim D_2$, $D_1 \neq D_2$. Let D_3 be a semi-reduced divisor with $D_3 \sim D_1 - D_2$ obtained as in the proof of Lemma 4.2. Since $D_1 \neq D_2$, there is a point P such that $\mathrm{ord}_P(D_1) \neq \mathrm{ord}_P(D_2)$. Suppose, without loss of generality, that $\mathrm{ord}_P(D_1) = m_1 \geq 1$, and either 1) $\mathrm{ord}_P(D_2) = 0$ and $\mathrm{ord}_{\widetilde{P}}(D_2) = 0$, or 2) $\mathrm{ord}_P(D_2) = m_2$ with $1 \leq m_2 < m_1$, or 3) $\mathrm{ord}_{\widetilde{P}}(D_2) = m_2$ with $1 \leq m_2 \leq m_1$. (If P is special, then 3) cannot occur.) In case 1), $\mathrm{ord}_P(D_3) = m_1 \geq 1$. In case 2), $\mathrm{ord}_P(D_3) = (m_1 - m_2) \geq 1$. In case 3), $\mathrm{ord}_P(D_3) = (m_1 + m_2) \geq 1$. In all cases, $\mathrm{ord}_P(D_3) \geq 1$, and so $D_3 \neq 0$. Also, $|D_3| \leq |D_1 - D_2| \leq |D_1| + |D_2| \leq 2g$. Let G be a nonzero rational function in $\overline{\mathbb{F}}(C)$ such that $\mathrm{div}(G) = D_3$; since $D_1 \sim D_2$, and $D_3 \sim D_1 - D_2$, we know that D_3 is principal and hence such a function G exists. By Lemma 6.1, since G has no finite poles, it must be a polynomial function. Then $G = a(u) - b(u)v$ for some $a, b \in \overline{\mathbb{F}}[u]$. Since $\deg(v) = 2g + 1$ and $\deg(G) = |D_3| \leq 2g$, we must have $b(u) = 0$. Suppose that $\deg_u(a(u)) \geq 1$, and let $x \in \overline{\mathbb{F}}$ be a root of $a(u)$. Let $P = (x, y)$ be a point on C. Now, if P is ordinary, then both P and $\widetilde{P}$ are zeros of G, contradicting the fact that D_3 is semi-reduced. If P is special, then it must also be a zero of G of order at least 2, again contradicting the fact that D_3 is semi-reduced. Thus, $\deg_u(a(u)) = 0$ and so $D_3 = 0$, a contradiction. $\square$

§7. Adding Reduced Divisors

Let C be a hyperelliptic curve of genus g defined over a finite field $\mathbb{F}$, and let $\mathbb{J}$ be the jacobian of C. Let $P = (x, y) \in C$, and let σ be an automorphism of $\overline{\mathbb{F}}$ over $\mathbb{F}$. Then $P^\sigma \overset{\mathrm{def}}{=} (x^\sigma, y^\sigma)$ is also a point on C.

Definition 7.1. A divisor $D = \sum m_P P$ is said to be *defined over* $\mathbb{F}$ if $D^\sigma \overset{\mathrm{def}}{=} \sum m_P P^\sigma$ is equal to D for all automorphisms σ of $\overline{\mathbb{F}}$ over $\mathbb{F}$.

A principal divisor is defined over $\mathbb{F}$ if and only if it is the divisor of a rational function that has coefficients in $\mathbb{F}$. The set $\mathbb{J}(\mathbb{F})$ of all divisor classes in $\mathbb{J}$ that have a representative that is defined over $\mathbb{F}$ is a subgroup of $\mathbb{J}$. Each element of $\mathbb{J}(\mathbb{F})$ has a unique representation as a reduced divisor $\mathrm{div}(a, b)$, where $a, b \in \mathbb{F}[u]$, $\deg_u a \leq g$, $\deg_u b < \deg_u a$; and hence $\mathbb{J}(\mathbb{F})$ is in fact a finite abelian group. This section presents an efficient algorithm for adding elements in this group.

Let $D_1 = \mathrm{div}(a_1, b_1)$ and $D_2 = \mathrm{div}(a_2, b_2)$ be two reduced divisors defined over $\mathbb{F}$ (that is, $a_1, a_2, b_1, b_2 \in \mathbb{F}[u]$). Algorithm 1 finds a semi-reduced divisor $D = \mathrm{div}(a, b)$ with $a, b \in \mathbb{F}[u]$, such that $D \sim D_1 + D_2$. Algorithm 2 reduces D to an equivalent reduced divisor D'. Notation: $b \bmod a$ denotes the remainder polynomial when b is divided by a.

Algorithms 1 and 2 were presented in [Koblitz 1989]. They generalize earlier algorithms in [Cantor 1987], in which it was assumed that $h(u) = 0$ and $\mathrm{char}(\mathbb{F}) \neq 2$.

Algorithm 1

INPUT: Semi-reduced divisors $D_1 = \mathrm{div}(a_1, b_1)$ and $D_2 = \mathrm{div}(a_2, b_2)$, both defined over $\mathbb{F}$.

OUTPUT: A semi-reduced divisor $D = \mathrm{div}(a, b)$ defined over $\mathbb{F}$ such that $D \sim D_1 + D_2$.

1) Use the Euclidean algorithm (see §3 of Chapter 3) to find polynomials d_1, e_1, $e_2 \in \mathbb{F}[u]$ where $d_1 = \mathrm{g.c.d.}(a_1, a_2)$ and $d_1 = e_1 a_1 + e_2 a_2$.
2) Use the Euclidean algorithm to find polynomials d, c_1, $c_2 \in \mathbb{F}[u]$ where $d = \mathrm{g.c.d.}(d_1, b_1 + b_2 + h)$ and $d = c_1 d_1 + c_2(b_1 + b_2 + h)$.
3) Let $s_1 = c_1 e_1$, $s_2 = c_1 e_2$, and $s_3 = c_2$, so that

$$d = s_1 a_1 + s_2 a_2 + s_3(b_1 + b_2 + h) . \tag{3}$$

4) Set

$$a = a_1 a_2 / d^2 \tag{4}$$

and

$$b = \frac{s_1 a_1 b_2 + s_2 a_2 b_1 + s_3(b_1 b_2 + f)}{d} \bmod a . \tag{5}$$

Theorem 7.1. *Let $D_1 = \mathrm{div}(a_1, b_1)$ and $D_2 = \mathrm{div}(a_2, b_2)$ be semi-reduced divisors. Let a and b be defined as in equations (4) and (5). Then $D = \mathrm{div}(a, b)$ is a semi-reduced divisor and $D \sim D_1 + D_2$.*

Proof. We first verify that b is a polynomial. Using equation (3), we can write

$$\frac{s_1 a_1 b_2 + s_2 a_2 b_1 + s_3(b_1 b_2 + f)}{d}$$

$$= \frac{b_2(d - s_2 a_2 - s_3(b_1 + b_2 + h)) + s_2 a_2 b_1 + s_3(b_1 b_2 + f)}{d}$$

$$= b_2 + \frac{s_2 a_2(b_1 - b_2) - s_3(b_2^2 + b_2 h - f)}{d} .$$

Since $d|a_2$ and $a_2|(b_2^2 + b_2h - f)$, b is indeed a polynomial.

Let $b = (s_1a_1b_2 + s_2a_2b_1 + s_3(b_1b_2 + f))/d + sa$, where $s \in \mathbb{F}[u]$. Now

$$b - v = \frac{s_1a_1b_2 + s_2a_2b_1 + s_3(b_1b_2 + f) - dv}{d} + sa$$

$$= \frac{s_1a_1b_2 + s_2a_2b_1 + s_3(b_1b_2 + f) - s_1a_1v - s_2a_2v - s_3(b_1 + b_2 + h)v}{d} + sa$$

$$= \frac{s_1a_1(b_2 - v) + s_2a_2(b_1 - v) + s_3(b_1 - v)(b_2 - v)}{d} + sa \ . \tag{6}$$

From (6) it is not hard to see that $a|b^2 + bh - f$. Namely, $b^2 + bh - f$ is obtained by multiplying the left side of (6) by its conjugate: $(b - v)(b + v + h) = b^2 + bh - f$. Thus, to see that $a|b^2 + bh - f$ it suffices to show that a_1a_2 divides the product of $\big(s_1a_1(b_2 - v) + s_2a_2(b_1 - v) + s_3(b_1 - v)(b_2 - v)\big)$ with its conjugate; and this follows because $a_1|b_1^2 + b_1h - f = (b_1 - v)(b_1 + v + h)$ and $a_2|b_2^2 + b_2h - f = (b_2 - v)(b_2 + v + h)$. Lemma 5.3 now implies that $\mathrm{div}(a, b)$ is a semi-reduced divisor.

We now prove that $D \sim D_1 + D_2$. There are two cases to consider.

1) Let $P = (x, y)$ be an ordinary point. There are two subcases to consider.

   a) Suppose that $\mathrm{ord}_P(D_1) = m_1$, $\mathrm{ord}_{\tilde{P}}(D_1) = 0$, $\mathrm{ord}_P(D_2) = m_2$, and $\mathrm{ord}_{\tilde{P}}(D_2) = 0$, where $m_1 \geq 0$, $m_2 \geq 0$. Now $\mathrm{ord}_P(a_1) = m_1$, $\mathrm{ord}_P(a_2) = m_2$, $\mathrm{ord}_P(b_1 - v) \geq m_1$, and $\mathrm{ord}_P(b_2 - v) \geq m_2$. If $m_1 = 0$ or $m_2 = 0$ (or both) then $\mathrm{ord}_P(d_1) = 0$, whence $\mathrm{ord}_P(d) = 0$ and $\mathrm{ord}_P(a) = m_1 + m_2$. If $m_1 \geq 1$ and $m_2 \geq 1$, then, since $(b_1 + b_2 + h)(x) = 2y + h(x) \neq 0$, we have $\mathrm{ord}_P(d) = 0$ and $\mathrm{ord}_P(a) = m_1 + m_2$. From equation (6) it follows that

   $$\mathrm{ord}_P(b - v) \geq \min\{m_1 + m_2, m_2 + m_1, m_1 + m_2\} = m_1 + m_2 \ .$$

   Hence, $\mathrm{ord}_P(D) = m_1 + m_2$.

   b) Suppose that $\mathrm{ord}_P(D_1) = m_1$ and $\mathrm{ord}_{\tilde{P}}(D_2) = m_2$, where $m_1 \geq m_2 \geq 1$. We have $\mathrm{ord}_P(a_1) = m_1$, $\mathrm{ord}_P(a_2) = m_2$, $\mathrm{ord}_P(d_1) = m_2$, $\mathrm{ord}_P(b_1 - v) \geq m_1$, $\mathrm{ord}_P(b_2 - v) = 0$, and $\mathrm{ord}_{\tilde{P}}(b_2 - v) \geq m_2$. The last inequality implies that $\mathrm{ord}_P(b_2 + h + v) \geq m_2$, and hence $\mathrm{ord}_P(b_1 + b_2 + h) \geq m_2$ or $(b_1 + b_2 + h) = 0$. It follows that $\mathrm{ord}_P(d) = m_2$ and $\mathrm{ord}_P(a) = m_1 - m_2$. From equation (6) it follows that

   $$\mathrm{ord}_P(b - v) \geq \min\{m_1 + 0, m_2 + m_1, m_1 + 0\} - m_2 = m_1 - m_2 \ .$$

   Hence, $\mathrm{ord}_P(D) = m_1 - m_2$.

2) Let $P = (x, y)$ be a special point. There are two subcases to consider.

   a) Suppose that $\mathrm{ord}_P(D_1) = 1$ and $\mathrm{ord}_P(D_2) = 1$. Then $\mathrm{ord}_P(a_1) = 2$, $\mathrm{ord}_P(a_2) = 2$, and $\mathrm{ord}_P(d_1) = 2$. Now $(b_1 + b_2 + h)(x) = 2y + h(x) = 0$, whence either $\mathrm{ord}_P(b_1 + b_2 + h) \geq 2$ or $b_1 + b_2 + h = 0$. It follows that $\mathrm{ord}_P(d) = 2$ and $\mathrm{ord}_P(a) = 0$. Hence, $\mathrm{ord}_P(D) = 0$.

   b) Suppose that $\mathrm{ord}_P(D_1) = 1$ and $\mathrm{ord}_P(D_2) = 0$. Then $\mathrm{ord}_P(a_1) = 2$, $\mathrm{ord}_P(a_2) = 0$, whence $\mathrm{ord}_P(d_1) = \mathrm{ord}_P(d) = 0$ and $\mathrm{ord}_P(a) = 2$. Since $\mathrm{ord}_P(b_1 - v) = 1$, it follows from equation (6) that $\mathrm{ord}_P(b - v) \geq 1$. It can be inferred from equation (6) that $\mathrm{ord}_P(b - v) \geq 2$ only if $\mathrm{ord}_P(s_2a_2 +$

$s_3(b_2 - v)) \geq 1$. If this is the case, then $\text{ord}_P(s_2a_2 + s_3(b_2 + h + v)) \geq 1$, and hence $\text{ord}_P(s_2a_2 + s_3(b_1 + b_2 + h)) \geq 1$ (or $s_2a_2 + s_3(b_1 + b_2 + h) = 0$). It now follows from equation (3) that $\text{ord}_P(d) \geq 1$, a contradiction. Hence $\text{ord}_P(b - v) = 1$, whence $\text{ord}_P(D) = 1$. $\square$

Example 7.1. Consider the hyperelliptic curve $C : v^2 + (u^2 + u)v = u^5 + u^3 + 1$ of genus $g = 2$ over the finite field $\mathbb{F}_{2^5}$ (see Example 1.3). $P = (\alpha^{30}, 0)$ is an ordinary point in $C(\mathbb{F}_{2^5})$, and the opposite of P is $\widetilde{P} = (\alpha^{30}, \alpha^{16})$. $Q_1 = (0, 1)$ and $Q_2 = (1, 1)$ are special points in $C(\mathbb{F}_{2^5})$. The following are examples of computing the semi-reduced divisor $D = \text{div}(a, b) = D_1 + D_2$, for sample reduced divisors D_1 and D_2 (see Algorithm 1).

1) Let $D_1 = P + Q_1 - 2\infty$ and $D_2 = \widetilde{P} + Q_2 - 2\infty$ be two reduced divisors. Then $D_1 = \text{div}(a_1, b_1)$, where $a_1 = u(u + \alpha^{30})$, $b_1 = \alpha u + 1$, and $D_2 = \text{div}(a_2, b_2)$, where $a_2 = (u + 1)(u + \alpha^{30})$, $b_2 = \alpha^{23}u + \alpha^{12}$.
   1) $d_1 = \text{g.c.d.}(a_1, a_2) = u + \alpha^{30}$; $d_1 = a_1 + a_2$.
   2) $d = \text{g.c.d.}(d_1, b_1 + b_2 + h) = u + \alpha^{30}$; $d = 1 \cdot d_1 + 0 \cdot (b_1 + b_2 + h)$.
   3) $d = a_1 + a_2 + 0 \cdot (b_1 + b_2 + h)$.
   4) Set $a = a_1a_2/d^2 = u(u + 1) = u^2 + u$, and

$$b = \frac{1 \cdot a_1b_2 + 1 \cdot a_2b_1 + 0 \cdot (b_1b_2 + f)}{d} \mod a$$

$$\equiv 1 \pmod{a} .$$

Check:

$$\text{div}(a) = 2Q_1 + 2Q_2 - 4\infty$$

$$\text{div}(b - v) = Q_1 + Q_2 + \sum_{i=1}^{3} P_i - 5\infty , \quad \text{where } P_i \neq Q_1, Q_2$$

$$\text{div}(a, b) = Q_1 + Q_2 - 2\infty .$$

2) Let $D_1 = P + Q_1 - 2\infty$ and $D_2 = Q_1 + Q_2 - 2\infty$. Then $D_1 = \text{div}(a_1, b_1)$, where $a_1 = u(u + \alpha^{30})$, $b_1 = \alpha u + 1$, and $D_2 = \text{div}(a_2, b_2)$, where $a_2 = u(u + 1)$, $b_2 = 1$.
   1) $d_1 = \text{g.c.d.}(a_1, a_2) = u$; $d_1 = \alpha^{14}a_1 + \alpha^{14}a_2$.
   2) $d = \text{g.c.d.}(d_1, b_1 + b_2 + h) = u$; $d = 1 \cdot u + 0 \cdot (b_1 + b_2 + h)$.
   3) $d = \alpha^{14}a_1 + \alpha^{14}a_2 + 0 \cdot (b_1 + b_2 + h)$.
   4) $a = (u + \alpha^{30})(u + 1)$; $b \equiv \alpha^{14}u + \alpha^{13} \pmod{a}$. Check:

$$\text{div}(a) = 2Q_2 + P + \widetilde{P} - 4\infty$$

$$\text{div}(b - v) = P + Q_2 + \sum_{i=1}^{3} P_i - 5\infty , \quad \text{where } P_i \neq P, \widetilde{P}, Q_2$$

$$\text{div}(a, b) = P + Q_2 - 2\infty .$$

3) Let $D_1 = P + Q_1 - 2\infty$ and $D_2 = P + Q_2 - 2\infty$. Then $D_1 = \text{div}(a_1, b_1)$, where $a_1 = u(u + \alpha^{30})$, $b_1 = \alpha u + 1$, and $D_2 = \text{div}(a_2, b_2)$, where $a_2 = (u + \alpha^{30})(u + 1)$, $b_2 = \alpha^{14}u + \alpha^{13}$.

1) $d_1 = \text{g.c.d.}(a_1, a_2) = (u + \alpha^{30})$; $d_1 = 1 \cdot a_1 + 1 \cdot a_2$.
2) $d = \text{g.c.d.}(d_1, b_1 + b_2 + h) = 1$.
3) $d = (\alpha^{15}u + \alpha^4)a_1 + (\alpha^{15}u + \alpha^4)a_2 + \alpha^{15} \cdot (b_1 + b_2 + h)$.
4) $a = u(u + 1)(u + \alpha^{30})^2$; $b \equiv \alpha^{17}u^3 + \alpha^{26}u^2 + \alpha^2 u + 1 \pmod{a}$. Check:

$$\text{div}(a) = 2P + 2\tilde{P} + 2Q_1 + 2Q_2 - 8\infty$$

$$\text{div}(b - v) = 2P + Q_1 + Q_2 + \sum_{i=1}^{2} P_i - 6\infty , \quad \text{where } P_i \neq P, \tilde{P}, Q_1, Q_2$$

$$\text{div}(a, b) = 2P + Q_1 + Q_2 - 4\infty .$$

Algorithm 2

INPUT: A semi-reduced divisor $D = \text{div}(a, b)$ defined over $\mathbb{F}$.
OUTPUT: The (unique) reduced divisor $D' = \text{div}(a', b')$ such that $D' \sim D$.
1) Set
$$a' = (f - bh - b^2)/a$$
and
$$b' = (-h - b) \bmod a' .$$

2) If $\deg_u a' > g$ then set $a \leftarrow a'$, $b \leftarrow b'$ and go to step 1.
3) Let c be the leading coefficient of a', and set $a' \leftarrow c^{-1}a'$.
4) Output(a', b').

Theorem 7.2. *Let $D = \text{div}(a, b)$ be a semi-reduced divisor. Then the divisor $D' = \text{div}(a', b')$ returned by Algorithm 2 is reduced, and $D' \sim D$.*

Proof. Let $a' = (f - bh - b^2)/a$ and $b' = (-h - b) \bmod a'$. We show that

1) $\deg_u(a') < \deg_u(a)$;
2) $D' = \text{div}(a', b')$ is semi-reduced; and
3) $D \sim D'$.

The theorem then follows by repeated application of the reduction process (step 1 of Algorithm 2).
1) Let $m = \deg_u a$, $n = \deg_u b$, where $m > n$ and $m \geq g + 1$. Then $\deg_u a' = \max(2g + 1, 2n) - m$. If $m > g + 1$, then $\max(2g + 1, 2n) \leq 2(m - 1)$, whence $\deg_u a' \leq m - 2 < \deg_u a$. If $m = g + 1$, then $\max(2g + 1, 2n) = 2g + 1$, whence $\deg_u a' = g < \deg_u a$.
2) Now $f - bh - b^2 = aa'$. Reducing both sides modulo a', we obtain

$$f + (b' + h)h - (b' + h)^2 \equiv 0 \pmod{a'} ,$$

which simplifies to

$$f - b'h - (b')^2 \equiv 0 \pmod{a'} .$$

Hence $a' | (f - b'h - (b')^2)$. It follows from Lemma 5.3 that $\text{div}(a', b')$ is semi-reduced.

3) Let $C_0 = \{P \in \text{supp}(D) : P \text{ is special}\}$, $C_1 = \{P \in \text{supp}(D) : P \text{ is ordinary}\}$, and $C_2 = \{\widetilde{P} : P \in C_1\}$, so that

$$D = \sum_{P_i \in C_0} P_i + \sum_{P_i \in C_1} m_i P_i - (*)\infty .$$

Then, as in the proof of Theorem 5.1, we can write

$$\text{div}(a) = \sum_{P_i \in C_0} 2P_i + \sum_{P_i \in C_1} m_i P_i + \sum_{P_i \in C_1} m_i \widetilde{P}_i - (*)\infty$$

and

$$\text{div}(b - v) = \sum_{P_i \in C_0} P_i + \sum_{P_i \in C_1} n_i P_i + \sum_{P_i \in C_1} 0\widetilde{P}_i + \sum_{P_i \in C_3} s_i P_i - (*)\infty ,$$

where $n_i \geq m_i$, C_3 is a set of points in $C \backslash (C_0 \cup C_1 \cup C_2 \cup \{\infty\})$, $s_i \geq 1$, and $s_i = 1$ if P_i is special. Since $b^2 + bh - f = N(b - v)$, it follows from Lemma 4.1 that

$$\text{div}(b^2 + bh - f)$$
$$= \sum_{P_i \in C_0} 2P_i + \sum_{P_i \in C_1} n_i P_i + \sum_{P_i \in C_1} n_i \widetilde{P}_i + \sum_{P_i \in C_3} s_i P_i + \sum_{P_i \in C_3} s_i \widetilde{P}_i - (*)\infty ,$$

and hence

$$\text{div}(a') = \text{div}(b^2 + bh - f) - \text{div}(a)$$
$$= \sum_{P_i \in C_1'} t_i P_i + \sum_{P_i \in C_1'} t_i \widetilde{P}_i + \sum_{P_i \in C_3} s_i P_i + \sum_{P_i \in C_3} s_i \widetilde{P}_i - (*)\infty ,$$

where $t_i = n_i - m_i$ and $C_1' = \{P_i \in C_1 : n_i > m_i\}$. Now $b' = -h - b + sa'$ for some $s \in \mathbb{F}[u]$. If $P_i = (x_i, y_i) \in C_1' \cup C_3$, then $b'(x_i) = -h(x_i) - b(x_i) + s(x_i)a'(x_i) = -h(x_i) - y_i$. Then, as in the proof of Theorem 5.1, it follows that

$$\text{div}(b' - v)$$
$$= \sum_{P_i \in C_1'} 0 P_i + \sum_{P_i \in C_1'} r_i \widetilde{P}_i + \sum_{P_i \in C_3} 0 P_i + \sum_{P_i \in C_3} w_i \widetilde{P}_i + \sum_{P_i \in C_4} z_i P_i - (*)\infty ,$$

where $r_i \geq t_i$, $w_i \geq s_i$, $w_i = 1$ if $P_i \in C_3$ is special, and C_4 is a set of points in $C \backslash (C_1' \cup C_3 \cup \{\infty\})$. Hence,

$$\text{div}(a', b') = \sum_{P_i \in C_1'} t_i \widetilde{P}_i + \sum_{P_i \in C_3} s_i \widetilde{P}_i - (*)\infty$$

$$\sim - \sum_{P_i \in C_1'} t_i P_i - \sum_{P_i \in C_3} s_i P_i + (*)\infty$$

$$= D - \text{div}(b - v) ,$$

whence $D \sim D'$. $\square$

Note that all of the computations in Algorithms 1 and 2 take place in the field $\mathbb{F}$ itself (and not in any proper extensions of $\mathbb{F}$). In Algorithm 1, if $\deg_u a_1 \leq g$ and $\deg_u a_2 \leq g$, then $\deg_u a \leq 2g$. In this case, Algorithm 2 requires at most $1 + \lceil g/2 \rceil$ iterations of step 1.

Example 7.2. Consider the hyperelliptic curve $C : v^2 + (u^2 + u)v = u^5 + u^3 + 1$ of genus $g = 2$ over the finite field $\mathbb{F}_{2^5}$ (see Examples 1.3 and 7.1). Consider the semi-reduced divisor $D = (0, 1) + (1, 1) + (\alpha^5, \alpha^{15}) - 3\infty$. Then $D = \text{div}(a, b)$, where

$$a(u) = u(u + 1)(u + \alpha^5) = u^3 + \alpha^2 u^2 + \alpha^5 u$$

and

$$b(u) = \alpha^{17} u^2 + \alpha^{17} u + 1 .$$

Algorithm 2 yields

$$a'(u) = u^2 + \alpha^{15} u + \alpha^{26} ,$$
$$b'(u) = \alpha^{23} u + \alpha^{21} .$$

Hence, $D \sim \text{div}(a', b') = (\alpha^{28}, \alpha^7) + (\alpha^{29}, 0) - 2\infty$.

Exercises

1. Verify that the curves C in Examples 1.2 and 1.3 have no singular points (except for ∞).

2. Let $R \in \mathbb{F}(C)$ be a non-zero rational function, and let $P \in C$. Prove that $\text{ord}_P(R)$ does not depend on the representation of R as a ratio of polynomial functions (see Definition 3.4).

3. Prove Lemma 5.3.

4. Let C be the curve in Example 1.2. Find the divisor of the polynomial function $G(u, v) = v^2 + uv + 6u^4 + 6u^3 + u^2 + 6u$.

5. Let C be the curve in Example 1.2. Find the polynomial representation for the semi-reduced divisor $D = 2(2, 2) + 3(5, 3) + (1, 1) + (6, 4)$.

6. Let C be the curve in Example 1.2. Use Algorithm 1 to compute $D_3 = \text{div}(a_3, b_3) = D_1 + D_2$, where $D_1 = \text{div}(u^2 + 6, 2u + 6)$ and $D_2 = \text{div}(u^2 + 4u + 2, 4u + 1)$. Check your work by computing these divisors explicitly.

7. Let C be the curve in Example 1.2. Consider the semi-reduced divisor $D = \text{div}(u^7 + 2u^6 + 3u^5 + 6u^3 + 4u + 5, 5u^6 + 5u^5 + 6u^4 + 4u^3 + 5u^2 + 4)$. Use Algorithm 2 to find the reduced divisor equivalent to D.

Answers to Exercises

Chapter 1

1. Prove that $m^{ed} \equiv 1 \pmod{p}$ and $\pmod{q}$ separately. When working modulo p, the case $m \equiv 0 \pmod{p}$ is trivial, and the case $m \not\equiv 0 \pmod{p}$ reduces to Fermat's Little Theorem, i.e., the congruence $m^{p-1} \equiv 1 \pmod{p}$.

2. (a) $g^{u_1} y^{u_2} = g^{s^{-1}(H+xr)} = g^k \pmod{p}$. (b) No one knows any way to find (r,s) without knowing k and x, that is, without either being Alice or finding discrete logs in $\mathbb{F}_p^*$.

Chapter 2

§1. 1. f 2. e 3. f 4. g 5. i (in fact, $(n+1)^n \asymp e n^n$) 6. e 7. a 8. a 9. e 10. f. 11. a 12. $O(m^2 n)$ 13. $O(m^2 \ln^2 n)$ (or else $O(m \ln n(m + \ln n))$, which has a more complicated but more "accurate" $g(m,n)$) 14. $O(m^n)$ 15. $O(m^2 n^2)$.

§2. 1. $k - l$ or $k - l + 1$ bits. 2. (a) $O(k + \ln n)$, (b) $O(\ln n)$, (c) $O(\ln n)$, (d) $O(2^k)$, (e) $O(n^2 \ln n)$, (f) $O(n)$.

3. By showing that f_n is the lower-left entry in the n-th power of the matrix $\begin{pmatrix} 0 & 1 \\ 1 & 1 \end{pmatrix}$ and then diagonalizing this matrix, derive the formula $f_n = = (\eta^n - \bar{\eta}^n)/\sqrt{5}$, where η is the golden ratio $(1 + \sqrt{5})/2$ and $\bar{\eta}$ is its conjugate $(1 - \sqrt{5})/2$. Then choose $g(n) = n \log_2 \eta \approx 0.694242n$.

5. $g(n) = (\log_2 e)n \ln n$. 6. 15000.

7. (c)<(e)<(d)<(a)<(b), since (a) $\asymp (\log_{10} 2)n \approx 0.3n$, (b) $\asymp n$, (c) $\asymp 5 \log_2 n$, (d) $\asymp \frac{1}{2}\sqrt{n} \log_2 n$, (e) $\asymp 2\sqrt{n}/\ln n$.

§3. 1. (a) $O(n^2)$, (b) $O(n^2 \ln^2 n)$, (c) $O(n^2 \ln^2 N)$. 2. 1000000.

3. (a) $O(n \ln^2 n)$, (b) $O(\ln^2 n)$. 4. (a) 16 hours, (b) 1000000 years.

5. (a) $O(n^2)$, (b) $O(n^4)$. 6. (a) $O(2^k)$, (b) $O(2^{2k})$, (c) $O(k2^k)$.

7. $O((k+l)^2)$ (also correct: $O(k^2 + l^2)$). 8. $O(l^2)$.

9. (a) $O(k^2 l^2)$, (b) $O(k^2 l^2)$, (c) $O(k^2 l^2)$, (d) $O(k^2 l)$, (e) $O(k^2 l^2)$, (f) $O(kl^2)$, (g) $O(k^2 l^2)$, (h) $O(K^2)$.

10. (c)<(d)<(b)<(e)<(a), since (a) $O(\ln^4 n)$, (b) $O(\ln^2 n)$, (c) negligible (since one just replaces each block of 4 bits by the corresponding name of a hexadecimal digit), (d) $O(\ln n(\ln \ln n)^2)$, (e) $O(\ln^3 n)$.

11. 0.00023 sec for $\gamma = 0$, 171 sec for $\gamma = 1/3$, 74 years for $\gamma = 1/2$, $> 10^{86}$ years for $\gamma = 1$.

§4. 1. (d),(c),(b),(a). 2. (a) and (c). 3. $O(B^2 \ln B)$.

4. Given a map, construct the graph whose vertices correspond to regions of the map and whose edges connect vertices whose corresponding regions have a common border. It is not hard to see that coloring maps is equivalent to coloring *planar* graphs.

5. First find the length k of a path of minimal distance, using binary search. After that you need to actually find a path of distance k. To do this, start with city #1. Set $i_1 = 1$. For $j = 2, 3, \ldots, m$ increase the distance for each $(1, j)$ by 1, and each time ask if the new Traveling Salesrep problem still has a path of length k. The first time the answer is "no", choose that value of j – call it i_2 – to be the city you go to from city #1. Then go through the same procedure with $i_1 = 1$ replaced by i_2 (and taking the cities in the order $j = 2, 3, \ldots, i_2 - 1, i_2 + 1, \ldots, m$). Continue in this way until you go to all of the cities.

6. Given an instance of $\mathcal{P}_2$, choose your two integers for $\mathcal{P}_1$ to be ad and bc. The answer to $\mathcal{P}_2$ is "yes" if and only if these two integers are equal.

7. Given an instance of $\mathcal{P}_2$, construct an instance of $\mathcal{P}_1$ by taking the cross-product of each pair of vectors in the input of $\mathcal{P}_2$.

8. Given an instance of $\mathcal{P}_1$, let the $p(X)$ in $\mathcal{P}_2$ be the derivative of the $p(X)$ in $\mathcal{P}_1$.

9. Use the algorithm for $\mathcal{P}_2$ to find the complete prime factorization of N: $N = p_1^{\alpha_1} p_2^{\alpha_2} \cdots p_r^{\alpha_r}$. Once you have this factorization, you can easily compute the Euler φ-function $\varphi(N) = (p_1^{\alpha_1} - p_1^{\alpha_1 - 1})(p_2^{\alpha_2} - p_2^{\alpha_2 - 1}) \cdots (p_r^{\alpha_r} - p_r^{\alpha_r - 1})$. Now use the Euclidean algorithm to determine whether g.c.d.$(e, \varphi(N)) = 1$ and, if it is, find integers d and y such that $ed - y\varphi(N) = 1$. This d is the desired output for $\mathcal{P}_1$. If g.c.d.$(e, \varphi(N))$ turns out to be greater than 1, then simply state that no such d exists.

10. Use the algorithm for $\mathcal{P}_2$ to find k and l in $\mathcal{P}_1$, after which it is trivial to find g^{kl} modulo p.

11. (a) No, because, as far as anyone knows, the only way to demonstrate the correctness of an answer is to go through the calculation of the exact value of $\pi(N)$, and no way is known to do this in polynomial time.

(b) No, because it is hard to imagine a certificate of the non-existence of paths of length less than or equal to k. This problem is the reverse of the Traveling Salesrep decision problem in the sense that the answer to the question in Exercise 11(b) is "yes" if and only if the answer to the Traveling Salesrep question is "no". Sometimes this problem is called "co-Traveling Salesrep". It belongs to the class co-NP, consisting of all problems whose co-problems are in NP. If this problem were in NP, it would follow that the class NP and the class co-NP are identical. It is conjectured that NP≠co-NP. As in the case of the P≠NP conjecture, mathematicians and computer scientists would be absolutely astounded if this conjecture were shown to be false. Thus, hardly anyone thinks that the co-Traveling Salesrep problem is in NP.

(c) No, because in general k might be exponentially large in the number of vertices, so a certificate of a "yes" answer cannot simply be a list of all k 3-colorings. It's hard to imagine what such a certificate could be.

12. False. When our NP problem $\mathcal{P}$ reduces to the NP-complete one, suppose that the input length is squared. Then an $L(1/2)$-algorithm for the NP-complete problem gives a fully exponential algorithm for $\mathcal{P}$.

§6. 1. $O(n^3)$, where $n = O(\ln N)$ is the input length.

2. For example, let $P_i = X^{2^{i-1}} + 1$. In that case $\prod_{i=1}^m P_i = \sum_{j=0}^{2^m - 1} X^j$.

§7. 1. Just look at the contribution to the sum in Definition 7.5 of the prime numbers N between 2^{n-1} and 2^n. This contribution is

$$\Omega\left(2^{-n}(\pi(2^n) - \pi(2^{n-1}))(\sqrt{2^n})^\varepsilon\right) = \Omega\left(\frac{1}{n}2^{\varepsilon n/2}\right),$$

where $\pi(2^n)$ denotes the number of primes less than 2^n.

2. Suppose that we are given any $\varepsilon > 0$. Take $\varepsilon' = \varepsilon/2$. The hypothesis (with ε' in place of ε) means that there exists $k > 1/\varepsilon'$ such that for $n \geq n_0$ the set S of instances i of input length $\leq n$ for which $T(i) \geq n^k$ is large enough so that $\sum_{i \in S} \mu_n(i) \geq n^{-k\varepsilon'}$. Then $\sum_{i \in S} T(i)^\varepsilon \mu_n(i) \geq n^{k(\varepsilon - \varepsilon')} = n^{k\varepsilon'}$, which is not $O(n)$, since $k\varepsilon' > 1$.

3. The implication in one direction can be proved directly or from Hölder's inequality. To disprove the converse, suppose, for example, that $T(i) = 4^n$ for $i \in S$ and $T(i) \leq n$ for $i \notin S$, where S is a set of instances such that $\sum_{i \in S} \mu_n(i) = 2^{-n}$. Then Levin's property holds, but the modified property does not.

4. Use a processor for each power of X. If n is the maximum degree of the two polynomials, then you have $O(n)$ processors, each of which takes $O(\ln n)$ bit operations, because of the condition on the size of the coefficients.

5. If a single processor did the work of the $O(n^{C_2})$ processors in series (i.e., one at a time), the time it would need is $O(\ln^{C_1} n^{C_2}) = O(n^{C_1 + C_2})$.

Chapter 3

§1. 1. (a) 6, since you need to adjoin both $\sqrt[3]{2}$ and $\sqrt{-3}$;
(b) 2, since you need only to adjoin $\sqrt{-3}$; (c) 2, much like part (b);
(d) 3, since $\mathbb{F}_7$ already has $\sqrt{-3} = \pm 2$, but you need to adjoin $\sqrt[3]{2}$;
(e) 1, since $\mathbb{F}_{31}$ already has 3 roots of the polynomial (namely: 4, 7, 20).
2. The criterion is that X^j occurs with nonzero coefficient only if $p|j$. In that case the polynomial is the p-th power of the polynomial obtained from it by replacing each X^j by $X^{j/p}$ (see Lemma 2.2).

§2. 1.

prime p	2	3	5	7	11	17
smallest generator	1	2	2	3	2	3
number of generators	1	1	2	2	4	8

2. (a) If $g^{p-1} \equiv 1 \bmod p^2$, then replace g by $(p+1)g$ and show that then one has $g^{p-1} = 1 + g_1 p$ with g_1 prime to p. Now if $g^j \equiv 1 \bmod p^\alpha$, first show that $p - 1 | j$, i.e., $j = (p-1)j_1$, and so $(1 + g_1 p)^{j_1} \equiv 1 \bmod p^\alpha$. But show that $(1 + g_1 p)^{j_1} = 1 + j_1 g_1 p +$ higher powers of p, and then $p^{\alpha-1}$ must divide j_1.

(b) For the first part, show that 1, $2^{\alpha-1} \pm 1$, and $2^\alpha - 1$ are all square roots of 1 modulo 2^α, and so the group is not cyclic; the proof of the second part (which reduces to showing that 5^j cannot be $\equiv 1 \bmod 2^\alpha$ unless $2^{\alpha-2} | j$) is similar to part (a).

3. You need the 7th roots of unity in order to have a splitting field; the degree f of the splitting field is the smallest power such that $p^f \equiv 1 \bmod 7$; this is either 1, 2, 3 or 6.

4. 2 for $d = 1$: X, $X + 1$; 1 for $d = 2$: $X^2 + X + 1$; 2 for $d = 3$: $X^3 + X^2 + 1$, $X^3 + X + 1$; 3 for $d = 4$: $X^4 + X^3 + 1$, $X^4 + X + 1$, $X^4 + X^3 + X^2 + X + 1$; 6 for $d = 5$: $X^5 + X^3 + 1$, $X^5 + X^2 + 1$, $X^5 + X^4 + X^3 + X^2 + 1$, $X^5 + X^4 + X^3 + X + 1$, $X^5 + X^4 + X^2 + X + 1$, $X^5 + X^3 + X^2 + X + 1$; 9 for $d = 6$: $X^6 + X^5 + 1$, $X^6 + X^3 + 1$, $X^6 + X + 1$, $X^6 + X^5 + X^4 + X^2 + 1$, $X^6 + X^5 + X^4 + X + 1$, $X^6 + X^5 + X^3 + X^2 + 1$, $X^6 + X^5 + X^2 + X + 1$, $X^6 + X^4 + X^3 + X + 1$, $X^6 + X^4 + X^2 + X + 1$.

5. 3 for $d = 1$: X, $X \pm 1$; 3 for $d = 2$: $X^2 + 1$, $X^2 \pm X - 1$; 8 for $d = 3$: $X^3 + X^2 \pm (X - 1)$, $X^3 - X^2 \pm (X + 1)$, $X^3 \pm (X^2 - 1)$, $X^3 - X \pm 1$; 18 for $d = 4$; 48 for $d = 5$; 116 for $d = 6$. 6. $(p^f - p^{f/\ell})/f$.

7. (a) Raising $0 = \alpha^2 + b\alpha + c$ to the p-th power and using the fact that $b^p = b$ and $c^p = c$, we obtain $0 = (\alpha^p)^2 + b\alpha^p + c$.

(b) The polynomial's two distinct roots are then α and α^p. Then a is minus the sum of the roots, and b is the product of the roots.

(c) $(c\alpha + d)^{p+1} = (c\alpha^p + d)(c\alpha + d)$, and then multiply out and use part (b).

(d) $(2 + 3i)^{5(19+1)+1} = (2^2 + 3^2)^5(2 + 3i) = 14(2 + 3i) = 9 + 4i$.

8. (a) Let α be a root of $X^2 + X + 1 = 0$; then the three successive powers of α are α, $\alpha + 1$, and 1.

(b) Let α be a root of $X^3 + X + 1 = 0$; then the seven successive powers of α are α, α^2, $\alpha + 1$, $\alpha^2 + \alpha$, $\alpha^2 + \alpha + 1$, $\alpha^2 + 1$, 1.

(c) Let α be a root of $X^3 - X - 1 = 0$; then the 26 successive powers of α are α, α^2, $\alpha + 1$, $\alpha^2 + \alpha$, $\alpha^2 + \alpha + 1$, $\alpha^2 - \alpha + 1$, $-\alpha^2 - \alpha + 1$, $-\alpha^2 - 1$, $-\alpha + 1$, $-\alpha^2 + \alpha$, $\alpha^2 - \alpha - 1$, $-\alpha^2 + 1$, -1, followed by the same 13 elements with all $+$'s and $-$'s reversed.

(d) Let α be a root of $X^2 - X + 2 = 0$; then the 24 successive powers of α are α, $\alpha - 2$, $-\alpha - 2$, $2\alpha + 2$, $-\alpha + 1$, 2, then the same six elements multiplied by 2, then multiplied by -1, then multiplied by -2, giving all 24 powers of α.

9. (a) $p = 2$ and $2^f - 1$ is a "Mersenne prime".

(b) Besides the cases in part (a), also you can have: (1) $p = 3$ and $(3^f - 1)/2$ a prime (as in part (a), this requires that f itself be prime, but that is not sufficient, as the example $f = 5$ shows), and (2) p of the form $2p' + 1$ with p' a prime and $f = 1$. It is not known, incidentally, whether there are infinitely many finite fields with any of the conditions in (a)–(b) (but it is conjectured that there are). Primes p' for which $p = 2p' + 1$ is also prime are called "Germain primes" after Sophie

Germain, who in 1823 proved that the first case of Fermat's Last Theorem holds if the exponent is such a prime.

10. Reduce to the case when $j = d$ by showing that $\sigma^j(a) = a$ and $\sigma^f(a) = a$ imply that $\sigma^d(a) = a$.

11. Show that $b' = b^{(p^f - 1)/(p^d - 1)}$ is in $\mathbb{F}_{p^d}$ by showing that it is fixed under σ^d (that is, raising to the p^d-th power); to show that it is a generator, note that all of the powers $(b')^j$, $j = 0, \ldots, p^d - 2$ are distinct, because the first $p^f - 1$ powers of b are distinct.

12. Let $d =$ g.c.d.$(k, p^f - 1)$. Since $d | p^f - 1$, the cyclic group $\mathbb{F}_{p^f}^*$ clearly has d d-th roots of unity. Each of them is also a k-th root. Conversely, by writing $d = uk + v(p^f - 1)$ you can show that any k-th root is also a d-th root.

13. For $x, x' \in \mathbb{K}$ it is easy to show that $g(x)^q = g(x)$ (and hence $g(x) \in \mathbb{F}_q$) and that $g(cx + c'x') = cg(x) + c'g(x')$ for $c, c' \in \mathbb{F}_q$. In order to show that $g(x)$ takes all possible values $y \in \mathbb{F}_q$, because of the $\mathbb{F}_q$-linearity of g it suffices to show that g is not identically zero. This follows because a polynomial of degree q^{n-1} cannot have q^n roots. The last assertion now follows because if V denotes the $(n-1)$-dimensional $\mathbb{F}_q$-subspace of $\mathbb{K}$ that g maps to 0, then $x_0 + V$ is the set of elements that g maps to $y_0 \in \mathbb{F}_q$ (here x_0 is any fixed element of $\mathbb{K}$ whose trace is y_0, and the notation $x_0 + V$ means all vectors x such that $x - x_0 \in V$).

14. (a) This follows immediately from the fact that $\mathbb{F}_q^*$ is cyclic.

(b) The only difference with the situation over $\mathbb{Z}$ is that, when dividing a by b (or r_{j-1} by r_j), one chooses the quotient to be the Gaussian integer that lies closest to a/b in the complex plane. (If there are two or more equally distant, then we choose one of them arbitrarily.) For example, $29 = 2(12 + i) + (5 - 2i)$. (c) The Gaussian integers have unique factorization (up to multiplication by the units ± 1, $\pm i$). The prime factorization of p is $(c + di)(c - di)$, where c and d are integers such that $c^2 + d^2 = p$. Since $p | (y + i)(y - i)$, it follows that either $c + di$ or $c - di$ must divide $y + i$, and hence must be the g.c.d. of p and $y + i$.

§3. 1. (a) $d(X) = 1 = X^2 g + (X + 1)f$; (b) $d(X) = X^3 + X^2 + 1 = f + (X^2 + X)g$; (c) $d(X) = 1 = (X - 1)f - (X^2 - X + 1)g$; (d) $d(X) = X + 1 = (X - 1)f - (X^3 - X^2 + 1)g$; (e) $d(X) = X + 78 = (50X + 20)f + (51X^3 + 26X^2 + 27X + 4)g$.

2. Since g.c.d.$(f, f') = X^2 + 1$, the multiple roots are $\pm \alpha^2$, where α is the generator of $\mathbb{F}_9^*$ in the text.

3. There exists a solution to a set of congruences modulo pairwise relatively prime polynomials, and that solution is unique up to multiples of the product of the moduli. In other words, there is a unique solution of degree less than the sum of the degrees of the moduli.

§4. 1. (a) the principal ideals generated by an irreducible polynomial; (b) the ideals generated by a prime number p and a polynomial $f \in \mathbb{Z}[X]$ whose reduction modulo p is irreducible as an element of $\mathbb{F}_p[X]$.

2. (a) $P_1 = (X)$, $P_2 = (X, Y)$; (b) $P_1 = (X)$, $P_2 = (X, p)$ where p is any prime; (c) $P_j = (X_1, \ldots, X_j)$ for $j = 1, \ldots, m$.

3. Show that the quotient ring R/I is a field of 3 elements. Show that if it were principal, then R/I would have to have more than 3 elements.

4. $I = (xy, y^2 - y)$.

5. Let I be the ideal consisting of all polynomials with zero constant term. Suppose that it is generated by a finite set $\{f_1, \ldots, f_m\}$. Let X_N be any variable not appearing in any of the f_i. Then the polynomial X_N is in I, but it cannot be written as a linear combination of the f_i.

6. Suppose that $f^n \in I$ and $g^m \in I$. Using the binomial expansion, show that $(f \pm g)^{n+m} \in I$. Thus, the radical is closed under addition and subtraction. The rest of the verification that the radical is an ideal is immediate.

§5. 1. When written as rows of a matrix, the coefficients of the linear forms must give a row-echelon matrix in (a) and a reduced row-echelon matrix in (b) (up to a rearrangement of the rows).

2. $l = 1$, and g_1 is the monic polynomial that generates the (principal) ideal I.

3. Since $g_1' \in I$, it follows that $\mathrm{lt}(g_1')$ is divisible by one of the $\mathrm{lt}(g_i)$, say $\mathrm{lt}(g_1)$. Similarly, $\mathrm{lt}(g_1)$ is divisible by one of the $\mathrm{lt}(g_i')$. But this i must be 1, because $\mathrm{lt}(g_i')$ cannot divide $\mathrm{lt}(g_1')$ for any $i \neq 1$, by the definition of a *minimal* Gröbner basis. Since both g_1 and g_1' are monic and $\mathrm{lt}(g_1)$ and $\mathrm{lt}(g_1')$ divide one another, it follows that g_1 and g_1' have the same leading term. Continue in this way for $g_2', g_3', \ldots$.

4. (a) False (see Exercise 6 below). (b) False (see Exercise 7 below). (c) True.

5. $\{X - Y, Y^2 - Y\}$. 6. The set of power products of total degree n.

7. $\{g_1, g_2, g_3, g_4, g_5\}$, where $g_4 = S(g_1, g_2) = Z^2$ and $g_5 = S(g_1, g_3) = X^2$. 8. $\{X - Y, Y^2 - Y\}$. 9. $\{g_1, g_2, g_3, g_4\}$, where $g_4 = S(g_1, g_3) = X^2 Z - Y^2 Z$.

10. $\{g_1, g_2, g_3, g_4\}$, where $g_4 = S(g_1, g_2) = -XY^2 + Y^3$, is a Gröbner basis; and $\{g_1, -g_4\} = \{X^2 - Y^2, XY^2 - Y^3\}$ is the reduced Gröbner basis.

11. Let $g_4 = S(g_1, g_3) = X^2 Y - Y^3$, $g_5 = S(g_2, g_3) = XY^2 - Y$, $g_6 = S(g_2, g_5) = -X^2 + Y^2$; then $\{g_1, \ldots, g_6\}$ is a Gröbner basis, and $\{g_2, g_5, -g_6\}$ is the reduced Gröbner basis.

12. Let $\mathbb{K} \subset \overline{\mathbb{F}}$ denote a finite extension of $\mathbb{F}$ that contains all of the coefficients of the polynomials that one is working with; and let $\beta_1 = 1, \beta_2, \ldots, \beta_l$ be a basis for $\mathbb{K}$ over $\mathbb{F}$. (a) In the relation that expresses f in terms of elements of I with coefficients in $\mathbb{K}[X]$, the coefficients (in $\overline{\mathbb{F}}$) of any power product can be equated, and so the β_1-component of those coefficients can be equated. The result is a relation expressing f in terms of elements of I with coefficients in $\mathbb{F}[X]$. (b) Similar to part (a). Show that if $f \in \overline{I}$, and if we write $f = \sum_{i=1}^{l} \beta_i f_i$, where $f_i \in \mathbb{F}[X]$, then each $f_i \in I$.

13. Use the previous exercise to reduce to the case where $\mathbb{F}$ is algebraically closed. If the set of points is empty, then $1 \in I$ by Theorem 4.2, and so $1 \in G$ and the claim is trivial. Otherwise, let $f \in \mathbb{F}[X_i]$ be a polynomial that vanishes at the i-th coordinates of all of the points. By Theorem 4.3, $f^n \in I$ for some n. Conclude that some power of X_i is divisible by the leading term of an element of G.

Chapter 4

§1. 1. Suppose that $a \geq b$. First show that if $a = qb + r$ is the first step of the Euclidean algorithm for a and b (i.e., r is the remainder when a is divided by b), then $q^r - 1$ is the remainder when $q^a - 1$ is divided by $q^b - 1$. Thus, the algorithm that gives g.c.d.$(q^a - 1, q^b - 1)$ mimics the algorithm that gives g.c.d.(a, b), in the sense that each remainder r_i in the computation of g.c.d.(a, b) becomes $q^{r_i} - 1$. See also Theorem 2.5 of Chapter 3, from which it follows that $\mathbb{F}_{q^d}$ is the largest field contained in both $\mathbb{F}_{q^a}$ and $\mathbb{F}_{q^b}$, where $d =$ g.c.d.(a, b).

2. Let $d =$ g.d.c.$(q^\theta + 1, q^n - 1)$. By Exercise 1, g.c.d.$(q^{2\theta} - 1, q^n - 1) = q - 1$. Since $q^\theta + 1$ divides $q^{2\theta} - 1$, this means that $d |$ g.c.d.$(q^\theta + 1, q - 1)$. Since $q^\theta + 1 \equiv 1^\theta + 1 = 2$ (mod $q - 1$), it follows that $d | 2$. But d must be odd, since q is even.

3. Let $n = n'\theta$, where n' is odd. Then $q^n - 1 \equiv (-1)^{n'} - 1 = -2$ (mod $q^\theta + 1$). Hence, g.c.d.$(q^\theta + 1, q^n - 1)$ divides 2. Again use the fact that q is even to conclude that $d = 1$.

4. $(0, 0, 0, 1, 1)$; $(0, 0, 1, 0, 1)$; $(1, 0, 0, 0, 0)$; $(0, 1, 0, 0, 0)$.

5. $(u_1 + u_2 X + u_3 X^2)^5 = (u_1^2 + u_2^2 + u_3^2 + u_2 u_3) + (u_2^2 + u_3^2 + u_1 u_3)X + (u_2^2 + u_1 u_2 + u_1 u_3)X^2$;
$v_1 = x_2 x_3 + x_3 + x_2 + x_3{}^2 + x_1{}^2 + x_1 + x_1 x_3$, $v_2 = x_2{}^2 + x_1{}^2 + x_2 x_3 + x_3 + x_2$,
$v_3 = x_1{}^2 + x_1 + x_2 x_1 + 1 + x_3{}^2$; $y_1 = x_2 x_3 + x_3 + x_2 + x_3{}^2 + x_1{}^2 + x_1 + x_1 x_3 + 1$,
$y_2 = x_3{}^2 + x_1 + x_1 x_3 + 1 + x_2{}^2$, $y_3 = x_1{}^2 + x_1 x_3 + x_2{}^2 + x_2 x_1$.

§2. 1. If 2 is replaced by a prime power $q > 2$ in (16), then for each nonzero solution $\mathbf{u}, \mathbf{v}$ we also have the solutions $\alpha \mathbf{u}, \mathbf{v}$ for nonzero $\alpha \in \mathbb{F}_q$. This contradicts the assumption that (15) gives a 1-to-1 correspondence between $\mathbf{u}$ and $\mathbf{v}$ and the fact that (15) and (16) are equivalent for nonzero $\mathbf{u}$ and $\mathbf{v}$.

2. Since $3h + 2 - (2^n - 1) = 2^{n-3}$, one has $\mathbf{v}^3 \mathbf{u}^2 = \mathbf{u}^{3h+2} = \mathbf{u}^{2^{n-3}}$.

§3. 1. The map $x \mapsto x^{2^{n-1}}$ inverts the squaring map; this proves bijectivity. When $q = 2$, the maps $\mathbf{v} = \mathbf{u}^2$ and $\mathbf{z} = \mathbf{w}^2$ are linear (see (2) with $q = 2$, $k = 1$). Hence $\overline{y} = D\overline{x}$ for some $n \times n$-matrix D. Such a cryptosystem can easily be broken by finding $O(n)$ plaintext/ciphertext pairs. If $q = 2^r$, the system is still easy to break, because $\mathbb{K}$ may be regarded as an extension of $\mathbb{F}_2$ of degree nr, in which case the ciphertext and plaintext (regarded as nr-tuples of 0's and 1's) are related by an $nr \times nr$-matrix.

2. No, since $\begin{pmatrix} 1 & \alpha \\ 0 & 1 \end{pmatrix}^2 = \begin{pmatrix} 1 & 0 \\ 0 & 1 \end{pmatrix}$ for all $\alpha \in \mathbb{F}_{2^n}$.

3. If i denotes a square root of -1 in $\mathbb{K}$, then $\pm \mathbf{u}$ and $\pm i \mathbf{u}$ all lead to the same $\overline{y}$.

4. It suffices to find a bilinear map $* : Y \times Y \to Y$ such that (20) holds, after which (21) holds (up to ± 1) with $h' = (q^n + 1)/4$; and C can be found as in §2.4 using $O(n)$ plaintext/ciphertext pairs. In order to find the bilinear map $*$, first find an n-dimensional space of matrices T such that for some matrix S one has $T \varphi(\overline{x}, \overline{x}') = \varphi(S\overline{x}, \overline{x}')$. Let T_i, $i = 1, \ldots, n$, be a basis for this space of matrices; find a matrix G by solving (26); and define $*$ by (25).

5. (a) The probability that none of the values is zero is equal to $\left(1 - \frac{1}{q}\right)^q \approx \frac{1}{e}$.
(b) Use the "inclusion–exclusion" counting principle to show that the probability

is

$$\binom{q}{1}\frac{1}{q} - \binom{q}{2}\frac{1}{q^2} + \binom{q}{3}\frac{1}{q^3} - \binom{q}{4}\frac{1}{q^4} + \cdots = 1 - \left(1 - \frac{1}{q}\right)^q .$$

Chapter 5

§2. 1. An integer n such that g^n is the identity but y^n is not. 2. If $n = 1$, $\mathcal{A}$ is always reversible, since it just translates the configuration; if n is even, $\mathcal{A}$ is never reversible, since $\mathcal{A}(C_1) = \mathcal{A}(C_0) = C_0$, where C_i is the configuration with all cells in state i.

§3. 1. (a) In Example 3.1, given a 3-coloring $v \mapsto i_v$, let $y_{v,i}$ take the value 1 if $i = i_v$ and 0 otherwise. In Example 3.2, given T_0, let y have t-coordinate 1 if $t \in T_0$ and 0 otherwise. In Example 3.2a, given $V' \subset V$, let y_v take the value 1 if $v \in V'$ and 0 otherwise. (b) In Subset Perfect Code, note that for each T_j and each $t \in T_j$ the ideal J contains the element $\left(\sum_{t' \in T_j, \ t' \neq t} tt'\right) + t\left(1 - \sum_{t' \in T_j} t'\right) = t - t^2$. The argument for 3-Coloring is similar.
2. $T = \{t_{v,i} : v \in V, 1 \leq i \leq m\}$; in B_1 take $t_{v,1} + t_{v,2} + \cdots + t_{v,m} - 1$, in B_2 let $1 \leq i < j \leq m$, and in B_3 let $1 \leq i \leq m$.
3. See 4.
4. Assume that $\mathbb{F}$ contains a primitive m-th root of unity ζ. (In particular, m is not divisible by the characteristic of $\mathbb{F}$.) Set $B_1' = \{x_v^m - 1 : v \in V\}$ and $B_2' = \{x_u^{m-1} + x_u^{m-2}x_v + x_u^{m-3}x_v^2 + \cdots + x_u^2 x_v^{m-3} + x_u x_v^{m-2} + x_v^{m-1} : uv \in E\}$.
(a) Given an m-coloring $v \mapsto i_v$, set y_v equal to ζ^{i_v}. (b) Set $x_v = \sum_{j=1}^m \zeta^j t_{v,j}$ and $t_{v,i} = \frac{1}{m}\sum_{j=1}^m (\zeta^{-i}x_v)^j$.
5. Obviously $J' \subset J''$. Suppose that $f \in J''$, i.e., $f(X,Y)$ vanishes at the six points $(x_i, y_i) = (1, \zeta)$, $(1, \bar{\zeta})$, $(\zeta, 1)$, $(\zeta, \bar{\zeta})$, $(\bar{\zeta}, 1)$, and $(\bar{\zeta}, \zeta)$. Modulo J' we can reduce $f(X,Y)$ to the form $aXY^2 + bXY + cY^2 + dX + eY + h$. Substituting $(X,Y) = (x_i, y_i)$ for $i = 1, 2, 3, 4, 5, 6$, we obtain six linear homogeneous equations in the six unknown coefficients a, b, c, d, e, h. Show that the determinant is nonzero, and hence $a = b = c = d = e = f = 0$.
6. Catherine reduces the ciphertext c modulo G' to get $c \sim m'$, where $\sim$ denotes "modulo the ideal J" and m' cannot be further reduced modulo G'. Since $c \sim m$, it follows that $m' - m \in J$. Note that m cannot be further reduced modulo G', since each $lt(g_i')$ is divisible by $lt(g_j)$ for some $g_j \in G$. Hence $m' - m = 0$.
7. (a) Modulo the ideal generated by B_1 (see Example 3.2a) one can write $b = \sum_{v \in V} c_v \equiv \sum_{v \in V} c_v \left(\sum_{u \in N[v]} t_u\right) = \sum_{u \in V} c_u' t_u$. (b) Regard the equations $c_v' = \sum_{u \in N[v]} c_u$ as a system of linear equations in the unknowns c_u.
8. (a) Same as 7(b). (c) In that case $m = \sum c_v = \frac{1}{r+1}\sum c_v'$.
9. In 3-Coloring start with a set of dots labeled 1, 2, or 3 at random. Draw random edges between pairs of vertices, never connecting two vertices with the same label. Then make another copy of the graph with the labels removed. In Perfect Code start with a set of dots that will be your solution V'; then draw several line segments emanating from each vertex. The outer endpoints of these

lines will be the vertices in $V \setminus V'$. Finally, draw a bunch of additional edges between the different outer endpoints; and make another copy of the graph that has no indication of the location of the original vertices.

10. (a) Let p_i correspond to t_i, let $\neg p_i$ correspond to $t_i - 1$, and let $\vee$ correspond to multiplication of polynomials. Let T correspond to 0 and F correspond to 1. Then any truth assignment that makes all of the clauses true corresponds to a function $\{t_i\} \longrightarrow \{0, 1\}$ that makes all of the corresponding polynomials vanish. It is easy to see that if a point at which all the polynomials vanish has a coordinate that is not 0 or 1, then that coordinate can be replaced by 0 or 1 without affecting the vanishing of the polynomials; in this way one gets a point that corresponds to a truth assignment map on the $\{p_i\}$. (b) Throw in the polynomials $t_i^2 - t_i$ for all i, thereby forcing the coordinates of points in the zero set to be 0 or 1.

11. Cathy chooses a random permutation π of the indices i, and for each i chooses an element $c_i'' \in J$. She then sets $c_i' = c_{\pi(i)} + c_{\pi(i)}''$. Here the c_i'' should be chosen so as to cancel many of the terms in c_i and change the degrees of some of the ciphertext polynomials, so that the set of ciphertext $\{c_i'\}$ does not look much like the set $\{c_i\}$. After Alice sends her the m_i', Cathy immediately finds $m_i = m_{\pi^{-1}(i)}'$.

§4. 1. Let p range through all prime numbers less than N, and for each p choose an irreducible polynomial $q_p(t_p)$ of degree p over $\mathbb{F}$, where $T = \{t_p\}_{p<N}$ is the set of variables. Let $c = 1$ in the Ideal Membership problem. The input length is proportional to $\sum p \approx N^2 / \ln N$. (See the Prime Number Theorem in §1 of Chapter 2. If you choose q_p to be sparse – namely, to have $O(\ln N)$ nonzero terms – then the input length is $O(N)$.) On the other hand, the extension degree of the field generated by a common zero $(\ldots, y_p, \ldots)$ is $\prod p$, which is of magnitude e^N.

Chapter 6

§1. 1. (a) Let $Y \to Y - (a_1 X + a_3)/2$ to get Y^2 on the left.
(b) Let $X \to X - a_3/a_1$. To reduce to the case when $a_1 = 1$, replace X by $a_1^2 X$ and Y by $a_1^3 Y$, and then divide the entire equation by a_1^6.

2. Let $f(X)$ be the cubic on the right in (1), and let x be any root of $f'(X) = X^2 + a_4$. Let y be a square root of $f(x)$. Then (x, y) is a point on the curve that satisfies (2).

3. (a) Let $f(X)$ be the cubic on the right. The Y-partial is zero at a point (x, y) when $y = 0$, i.e., $f(x) = 0$; and the X-partial is zero when $f'(x) = 0$. But $f(x) = f'(x) = 0$ has a solution if and only if $f(X)$ has a multiple root.
(b) If $a_1 = 0$ and $a_3 \neq 0$, then it is always smooth, because the Y-partial is the constant a_3. If $a_1 \neq 0$ and $a_3 = 0$, then it is smooth unless $a_6 = (a_4/a_1)^2$. If $a_1 \neq 0$, then to get rid of a_4 replace Y by $Y + a_4/a_1$; if $a_3 \neq 0$, then to get rid of a_2 replace Y by $Y + cX$ where $c^2 = a_2$.

4. You want to prove that $(P+Q)+R = P+(Q+R)$. Take the general case (where P, Q, and R are not collinear, are not negatives of one another, and are not the point at infinity). Let l_1 be the line through P, Q, and $-(P+Q)$; let l_2 be the line

through O, R, and $-R$; and let l_3 be the line through $-P$, $-(Q+R)$, and a third point $S = P+(Q+R)$. Let l'_1 be the line through Q, R and $-(Q+R)$; let l'_2 be the line through O, P, and $-P$; and let l'_3 be the line through $-R$, $-(P+Q)$ and a third point $S' = (P+Q)+R$. Conclude that $S = S'$, i.e., $P+(Q+R) = (P+Q)+R$.

5. Over $\mathbb{C}$ there are always n^2 points P such that $nP = O$; over $\mathbb{R}$ there are n when n is odd, and there are either n or $2n$ when n is even, depending on whether the curve has 1 or 2 connected components, respectively. (The curve $Y^2 = X^3 - X$ is an example with 2 connected components, and the curve $Y^2 = X^3 + X$ is an example with 1.)

6. (a) P is on the x-axis; (b) P is an inflection point; (c) P is a point where a line from an x-intercept of the curve is tangent to the curve.

7. $P + Q = (6,0)$, $2P = (\frac{25}{4}, -\frac{35}{8})$. 8. (a) 3; (b) 4; (c) 7.

9. In order to work only with integers, we can use projective coordinates (X, Y, Z) (also called "homogeneous coordinates"). Given a rational point (x, y), choose projective coordinates that are relatively prime integers (this determines those coordinates up to ± 1). Then instead of a bound on the denominator of the x-coordinate of a point, it suffices to find a bound on the maximum projective coordinate. When the equations (5) for doubling a point are written in terms of projective coordinates, one obtains X_3, Y_3, Z_3 as fourth degree polynomials in X_1, Y_1, Z_1. This gives a bound of the form $O(4^k)$ for the logarithm of the maximum of the projective coordinates of $2^k P$. To put it another way, the denominator of the x-coordinate of nP might grow as rapidly as $e^{O(n^2)}$ (here $n = 2^k$).

10. Use Exercise 3 to prove smoothness. To show that $N_1 = q+1$ in (a), note that if $x \neq 0, \pm 1$, then for exactly one of the pair $\pm x$ the expression $x^3 - x$ will have two square roots in $\mathbb{F}_q$ (this is because -1 is a non-square in $\mathbb{F}_q^*$); in (b) note that any element of $\mathbb{F}_q$ – in particular, $y^2 + y$ for any y – has exactly one cube root. Finally,

$$N_r = \begin{cases} q^r + 1, & r \text{ odd}; \\ q^r + 1 - 2(-q)^{r/2}, & r \text{ even}. \end{cases}$$

11. (a) If $P = (x, y)$, then $-P = (x, y+1)$ and $2P = (x^4, y^4 + 1)$.
(b) Use part (a) to find that $4P = (x^{16}, y^{16}) = (x, y) = P$.
(c) By part (b), we have $2P = -P$, i.e., $(x^4, y^4 + 1) = (x, y + 1)$; but this means that $x^4 = x$ and $y^4 = y$, so that $x, y \in \mathbb{F}_4$. By Hasse's theorem, the number N of points is within $2\sqrt{4} = 4$ of 5 and is within $2\sqrt{16} = 8$ of 17; hence, $N = 9$.

12. Both over $\mathbb{F}_2$ and $\mathbb{F}_3$ there is no solution to the equation, so the only point is the point at infinity. The numerator of the zeta-function is $1 - 2T + 2T^2$ and $1 - 3T + 3T^2$, respectively. N_r is the square of the complex absolute value of $(1 + i)^r - 1$ and $(1 + \omega)^r - 1$, respectively, where $\omega = (-1 + i\sqrt{3})/2$.

13. $Y^2 + Y = X^3 + \alpha$, where $\alpha \in \mathbb{F}_4$, $\alpha \neq 0, 1$. $N_r = (2^r - 1)^2$. Finally, for $P = (x, y)$ we have $2P = (x^4, y^4)$, and so $2^r P = (x^{4^r}, y^{4^r}) = (x, y) = P$.

14. See Example 3.5 of Chapter 2 (modular exponentiation).

15. Let $a_0 = 2$, $a_1 = a$, and $a_r = \alpha^r + \overline{\alpha}^r$. Then $N_r = q^r + 1 - a_r$ and $a_{r+1} = aa_r - qa_{r-1}$.

16. Let $N_{a,r} = \#E_a(\mathbb{F}_{2^r})$. Then $N_{a,1} = 3+(-1)^a$ and $Z(E_a/\mathbb{F}_2;T) = (1+(-1)^aT+2T^2)/(1-T)(1-2T)$. We have: $N_{0,5} = 4\cdot 11$, $N_{0,7} = 4\cdot 29$, $N_{0,13} = 4\cdot 2003$; $N_{1,3} = 2\cdot 7$, $N_{1,5} = 2\cdot 11$, $N_{1,7} = 2\cdot 71$, $N_{1,11} = 2\cdot 991$.

17. A point of order 2 exists on a curve in the form (1) if and only if $y = -a_1x - a_3 - y$ for some (x,y) on the curve.

§2. 1. (a) If x is Alice's secret key, then $u_1P + u_2Q = (u_1 + u_2x)P = s^{-1}(H(m) + xr)P = kP$. (b) No one knows any way to find (r,s) without knowing k and x, that is, without either being Alice or finding discrete logs on the elliptic curve. See also Exercise 2 at the end of Chapter 1.

2. In the case of the elliptic curve $Y^2 = f(X) = X^3 - X$ over $\mathbb{F}_q$ with $q \equiv 3$ (mod 4), let $x \in \mathbb{F}_q$ correspond to the message m. Note that precisely one of the pair $f(x), -f(x)$ is a square in $\mathbb{F}_q$. In some convenient way choose a subset $S \subset \mathbb{F}_q^*$, $\#S = (q-1)/2$, such that exactly one of the pair $\pm y$ belongs to S for each $y \in \mathbb{F}_q^*$. If $f(x) = 0$, imbed m as the point $(x,0)$. Otherwise, imbed m as the point (x,y) if $f(x)$ is a square and as the point $(-x,-y)$ if $f(x)$ is not a square, where y is chosen to be the unique square root of $f(x)$ (or $-f(x) = f(-x)$) that is in S. In the case of the curve $Y^2 + Y = X^3$, let $y \in \mathbb{F}_q$ correspond to the message m. If $y = 0$ or -1, imbed m as the point $(0,y)$; otherwise set $x = (y^2 + y)^{(2-q)/3}$, in which case (x,y) is a point on E. In the case of an arbitrary elliptic curve, choose a subset $S_0 \subset \mathbb{F}_q$ and a small subset $S_1 \subset \mathbb{F}_q$ such that every element of $\mathbb{F}_q$ can be written in at most one way as a sum of an element of S_0 and an element of S_1. For example, if $q = p$ and $\mathbb{F}_q$ is a prime field, we might choose S_1 to consist of the integers $0, 1 \ldots, 2^k - 1$ and S_0 to consist of $0, 2^k, 2\cdot 2^k, 3\cdot 2^k, \ldots, ([2^{-k}p] - 1)\, 2^k$. We let a message correspond to an element $x_0 \in S_0$, and then add x_1 to x_0 for various $x_1 \in S_1$ until we obtain a value $x = x_0 + x_1$ such that $f(x)$ is a square in $\mathbb{F}_q$. At that point we can use a probabilistic algorithm to find a square root y of $f(x)$ (see §1.8), and we can imbed m as the point (x,y).

3. (a) If u is such a solution, then $\mathrm{Tr}(z) = \mathrm{Tr}(u^2) + \mathrm{Tr}(u) = 2\mathrm{Tr}(u) = 0$, and so all z for which a solution u exists must have trace zero. Since the map $u \mapsto u^2 + u$ is 2-to-1, its image consists of half of $\mathbb{F}_q$; hence, the image consists of all z having trace zero. Now let $\bar{z} = \{\varepsilon_1, \ldots, \varepsilon_r\} \in \mathbb{F}_2^r$ be the vector obtained by expressing z in terms of the basis; and let $\bar{u} = \{\eta_1, \ldots, \eta_r\}$ be the unknown vector corresponding to a solution u of the equation $u^2 + u = z$. Let M be the matrix (with respect to the basis $\{\beta_1, \ldots, \beta_r\}$) of the squaring map, which is an $\mathbb{F}_2$-linear map on $\mathbb{F}_q$. Then the equation $u^2 + u = z$ is equivalent to the equation $(M + I)\bar{u} = \bar{z}$, where I is the $r \times r$ identity matrix. This equation can be solved by Gaussian elimination over $\mathbb{F}_2$.

(b) $\mathrm{Tr}(z) = 0$ if and only if an even number of components ε_i are 1. To find u, set $\eta_1 = 0$ and $\eta_i = \varepsilon_i + \eta_{i-1}$ for $i = 2, 3, \ldots, r$. That gives one of the solutions of the equation $u^2 + u = z$; the components of the other solution $u + 1$ are obtained by replacing η_i by $1 + \eta_i$, $i = 1, \ldots, r$.

(c) Let x be a random nonzero element of $\mathbb{F}_q$. In the case of equation (12) set $z = x + a_2 + x^{-2}a_6$, and compute the trace of z. If $\mathrm{Tr}(z) = 1$, choose a different x. If $\mathrm{Tr}(z) = 0$, then find a solution u to the equation $u^2 + u = z$ as in part (a). Set $y = xu$.

Then $y^2 + xy = x^2(u^2 + u) = x^2z = x^3 + a_2x^2 + a_6$, so that $P = (x, y)$ is a point on the curve (12). In the case of equation (13) set $z = a_3^{-2}(x^3 + a_4x + a_6)$, and compute the trace of z. If $\text{Tr}(z) = 1$, choose a different x. If $\text{Tr}(z) = 0$, then find u such that $u^2 + u = z$, and set $y = a_3u$. Then $y^2 + a_3y = a_3^2(u^2 + u) = a_3^2z = x^3 + a_4x + a_6$, as desired.

4. For any elliptic curve (12) over $\mathbb{F}_q$, the number $N = \#E(\mathbb{F}_q)$ is an even number in the interval $q + 1 - 2\sqrt{q} < N < q + 1 + 2\sqrt{q}$. For each such N, use Exercise 15 of §1 to rapidly compute the corresponding N_f. Test N_f/N for primality, and stop when you find N such that N_f/N is prime. It then remains to find $a_2, a_6 \in \mathbb{F}_q$ such that there are N points on the elliptic curve over $\mathbb{F}_q$ that is given by (12). Let a_2 run through $\mathbb{F}_q$ and a_6 run through $\mathbb{F}_q^*$. Since q is small, for each pair $a_2, a_6 \in \mathbb{F}_q$ the following simple algorithm to compute $N = \#E(\mathbb{F}_q)$ is fast enough. First set $N = 2$ (to account for the point at infinity and the point with zero x-coordinate). Then let x run through $\mathbb{F}_q^*$. For each such x set $z = x + a_2 + x^{-2}a_6$, and compute $\text{Tr}(z)$. If $\text{Tr}(z) = 0$, increment N by 2; if $\text{Tr}(z) = 1$, leave N unchanged.

5. $O\left(((\ln q^k)^{1/3}\right) \geq O\left(\left(((\ln q)^2 \ln q\right)^{1/3}\right) = O(\ln q)$.

6. Much as in Exercise 11 of §3 of Chapter 5, Cathy permutes the message units and then for each ciphertext $c = (lQ, M + l(k_AQ))$ she randomly chooses l' and sets $c' = (lQ + l'Q, M + l(k_AQ) + l'(k_AQ)) = ((l + l')Q, M + (l + l')(k_AQ))$. The tragically gullible Alice deciphers c' for Cathy, because she doesn't recognize any connection with the ciphertext she received from Bob.

§3. 1. How often is $(p+1)/4$ prime as p ranges over primes congruent to 3 modulo 4? How often is $(p+1)/6$ prime as p ranges over odd primes congruent to 2 modulo 3?

2. If you compute $nP \in E(\mathbb{Q})$ for $n = 1, 2, \ldots$, you find that $11P$ has denominator divisible by $p = 23$. This means that, if you work modulo 23, you find that 11 times the point $(0, 0)$ on $E(\mathbb{F}_{23})$ is equal to the point at infinity. Use Hasse's theorem to conclude that $(0, 0)$ does not generate the group $E(\mathbb{F}_{23})$.

3. For what prime values of r is $(1 + i)^r - 1$ a prime Gaussian integer?

§5. 1. Use part 3) of Theorem 5.1 and the fact that $|\alpha_i^r| = q^{r/2}$.

2. In part 3) of Theorem 5.1, group together the α_i according to which irreducible factor $(1 - \alpha_iT)$ divides.

3. Since $p|a_1$ and $p|a_2$, it follows that γ_1 and γ_2 are of the form $(ap \pm \sqrt{bp})/2$. Since $\alpha_i = (\gamma_i + \sqrt{\gamma_i^2 - 4p})/2$, it follows that when the product $N_r = (1 - \alpha_1^r)(1 - \overline{\alpha}_1^r)(1 - \alpha_2^r)(1 - \overline{\alpha}_2^r)$ is multiplied out, all of the terms but the 1 will be divisible by p.

4. Besides the point at infinity, solutions (x, y) of the equation come in pairs (x, y), $(x, y+1)$; hence, $M_1 \equiv 1 \pmod 2$. Show that $M_2 - 1$ is equal to twice the number of $x \in \mathbb{F}_4$ such that $f(x) \in \mathbb{F}_2$, and that the latter set is either $\mathbb{F}_2$ or all of $\mathbb{F}_4$. The last assertion is proved in a manner similar to Exercise 3.

5. Besides the point at infinity and the one point of the form $(0, y)$, the other points can be handled as in Exercise 4. If $g = 2$, show that a_1 is odd, a_2 is even, and hence $N_1 = P(1)$ is even. Since $N_1|N_r$, N_r is also even for any r.

6. Let $s_0 = 2$, $s_1 = \gamma_1$, $s_r = \alpha_1^r + \overline{\alpha}_1^r$, $t_0 = 2$, $t_1 = \gamma_2$, $t_r = \alpha_2^r + \overline{\alpha}_2^r$. Show that $N_r = (q^r+1)(q^r+1-(s_r+t_r))+s_r t_r$ and $s_{r+1} = \gamma_1 s_r - q s_{r-1}$, $t_{r+1} = \gamma_2 t_r - q t_{r-1}$.

§6. 1. $11 = ((-2)^5 - 1)/(-2 - 1)$, so choose $a = -2$; $J(\chi, \chi) = \zeta^2(-2 - \zeta)(-2 - \zeta^3) = 2 + \zeta + 4\zeta^2 + 2\zeta^3$. 2. Here is the table:

$a \bmod 7$	0	2	3	4	5	6
$\pm\zeta^k$	ζ^4	$-\zeta^3$	$-\zeta^5$	ζ	$-\zeta^6$	ζ^2

Appendix

4. $(1, 1) + (1, 5) + 2(2, 2) + 2(2, 3) + 4(6, 4) - 10\infty$. 5. $\mathrm{Div}(u^7 + 2u^6 + 5u^5 + 3u^4 + 3u^3 + 6u^2 + 5u + 3, 6u^6 + 4u^5 + 6u^4 + 2u^3 + 5u^2 + 3u + 3)$.
6. $D_3 = \mathrm{Div}(u^2 + 6u + 5, 4u + 1)$. 7. $\mathrm{Div}(u^2 + u + 5, 4u + 4)$.

Bibliography

W.W. Adams, P. Loustaunau (1994): *An Introduction to Gröbner Bases*, Amer. Math. Society, Providence.

L.M. Adleman (1979): A subexponential algorithm for the discrete logarithm problem with applications to cryptography, *Proc. 20th IEEE Symp. Foundations of Computer Science*, 55–60.

L.M. Adleman, J. DeMarrais (1993): A subexponential algorithm for discrete logarithms over all finite fields, *Math. Comp.* **61**, 1–15.

L.M. Adleman, J. DeMarrais, M.-D. Huang (1994): A subexponential algorithm for discrete logarithms over the rational subgroup of the jacobians of large genus hyperelliptic curves over finite fields, *Algorithmic Number Theory*, Lect. Notes Comp. Sci. **877**, Springer-Verlag, 28–40.

L.M. Adleman, M.-D. Huang (1992): *Primality Testing and Abelian Varieties over Finite Fields*, Lect. Notes Math. **1512**, Springer-Verlag.

L.M. Adleman, M.D. Huang (1996): Counting rational points on curves and abelian varieties over finite fields, in Henri Cohen, ed., *Algorithmic Number Theory, Proc. Second Intern. Symp., ANTS-II*, Springer-Verlag, 1–16.

L.M. Adleman, C. Pomerance, R.S. Rumely (1983): On distinguishing prime numbers from composite numbers, *Annals Math.* **117**, 173–206.

G.B. Agnew, R.C. Mullin, S.A. Vanstone (1993): An implementation of elliptic curve cryptosystems over $\mathbb{F}_{2^{155}}$, *IEEE Journal on Selected Areas in Communications* **11**, 804–813.

A.O.L. Atkin (1991): The number of points on an elliptic curve modulo a prime, unpublished manuscript.

E. Bach (1990): Number-theoretic algorithms, *Annual Reviews in Computer Science* **4**, 112–172.

E. Bach, J. Shallit (1996): *Algorithmic Number Theory*, Vol. 1, MIT Press.

R. Balasubramanian, N. Koblitz (1998): The improbability that an elliptic curve has subexponential discrete log problem under the Menezes-Okamoto-Vanstone algorithm, *J. Cryptology* **11**, 141–145.

T. Becker, V. Weispfenning (1993): *Gröbner Bases: A Computational Approach to Commutative Algebra*, Springer-Verlag.

S. Ben-David, B. Chor, O. Goldreich, M. Luby (1989): On the theory of average case complexity, *Proc. 21st ACM Symp. Theory of Computing*, 204–216.

E.R. Berlekamp (1970): Factoring polynomials over large finite fields, *Math. Comp.* **24**, 713–735.

L. Bertrand (1995): Computing a hyperelliptic integral using arithmetic in the Jacobian of the curve, *Applicable Algebra in Engineering, Communication and Computing* **6**, 275–298.

B.J. Birch, H.P.F. Swinnerton-Dyer (1963, 1965): Notes on elliptic curves I and II, *J. Reine Angew. Math.* **212**, 7–25 and **218**, 79–108.

I. Blake, X. Gao, A. Menezes, R.C. Mullin, S.A. Vanstone, T. Yaghoobian (1992): *Applications of Finite Fields*, Kluwer Acad. Publ.

D. Boneh, R. Lipton (1996): Algorithms for black-box fields and their applications to cryptography, *Advances in Cryptology – Crypto '96*, Springer-Verlag, 283–297.

G. Brassard (1979): A note on the complexity of cryptography, *IEEE Trans. Information Theory* **25**, 232–233.

E.F. Brickell (1985): Breaking iterated knapsacks, *Advances in Cryptology – Crypto '84*, Spinger-Verlag, 342–358.

E.F. Brickell, A.M. Odlyzko (1988): Cryptanalysis: A survey of recent results, *Proc. IEEE* **76**, 578–593.

D. Le Brigand (1991): Decoding of codes on hyperelliptic curves, *Eurocode '90*, Lect. Notes Comp. Sci. **514**, Springer-Verlag, 126–134.

J. Brillhart (1972): Note on representing a prime as a sum of two squares, *Math. Comp.* **26**, 1011–1013.

J. Brillhart, D.H. Lehmer, J.L. Selfridge, B. Tuckerman, S.S. Wagstaff, Jr. (1988): *Factorizations of $b^n \pm 1$, $b = 2, 3, 5, 6, 7, 10, 11, 12$ Up to High Powers*, Amer. Math. Soc.

J. Buchmann, V. Müller (1991): Computing the number of points of elliptic curves over finite fields, presented at Intern. Symp. on Symbolic and Algebraic Computation, Bonn, July 1991.

J. Buchmann, H.C. Williams (1987): On principal ideal testing in algebraic number fields, *J. Symbolic Comp.* **4**, 11-19.

J. Buchmann, H.C. Williams (1988): A key exchange system based on imaginary quadratic fields, *J. Cryptology* **1**, 107–118.

J. Buchmann, R. Scheidler, H.C. Williams (1994): A key-exchange protocol using real quadratic fields, *J. Cryptology* **7**, 171–199.

J. Buhler, N. Koblitz (1998): Lattice basis reduction, Jacobi sums, and hyperelliptic cryptosystems, *Bull. Austral. Math. Soc.* **57**, 147–154.

L. Caniglia, A. Galligo, J. Heintz (1988): Borne simple exponentielle pour les degrés dans le théorème des zéros sur un corps de caractéristique quelconque, *C. R. Acad. Sci. Paris* **307**, 255–258.

D. Cantor (1987): Computing in the jacobian of a hyperelliptic curve, *Math. Comp.* **48**, 95–101.

J.W.S. Cassels (1966): Diophantine equations with special reference to elliptic curves, *J. London Math. Soc.* **41**, 193–291.

J.W.S. Cassels, E.V. Flynn (1996): *Prolegomena to a Middlebrow Arithmetic of Curves of Genus 2*, Cambridge Univ. Press.

L. Charlap, D. Robbins (1988): An elementary introduction to elliptic curves I and II, *CRD Expository Reports* No. 31 and 34, Institute for Defense Analysis, Princeton.

H. Cohen (1993): *A Course in Computational Algebraic Number Theory*, Springer-Verlag.

D. Coppersmith (1984): Fast evaluation of logarithms in fields of characteristic two, *IEEE Trans. Information Theory* **30**, 587–594.

D. Coppersmith, A.M. Odlyzko, R. Schroeppel (1986): Discrete logarithms in $GF(p)$, *Algorithmica* **1**, 1–15.

J.-M. Couveignes (1994): Quelques calcules en théorie des nombres, Thesis, Université de Bordeaux I.

D.A. Cox, J. Little, D. O'Shea (1997): *Ideals, Varieties, and Algorithms: An Introduction to Computational Algebraic Geometry and Commutative Algebra*, 2nd ed., Springer-Verlag.

T. Denny, O. Schirokauer, D. Weber (1996): Discrete logarithms: the effectiveness of the index calculus method, in Henri Cohen, ed., *Algorithmic Number Theory, Proc. Second Intern. Symp., ANTS-II*, Springer-Verlag, 337–361.

A. Dickenstein, N. Fitchas, M. Giusti, C. Sessa (1991): The membership problem for unmixed polynomial ideals is solvable in single exponential time, *Discrete Appl. Math.* **33**, 73–94.

M. Dickerson (1989): The functional decomposition of polynomials, Ph.D. Thesis, Department of Computer Science, Cornell University.

L.E. Dickson (1952): *History of the Theory of Numbers. Volume 2. Diophantine Analysis*, Chelsea.

W. Diffie, H. Fell (1986): Analysis of a public key approach based on polynomial substitutions, *Advances in Cryptology – Crypto '85*, Springer-Verlag, 340–349.

W. Diffie, M.E. Hellman (1976): New directions in cryptography, *IEEE Trans. Information Theory* **22**, 644–654.

Do Long Van, A. Jeyanthi, R. Siromoney, K.G. Subramanian (1988): Public key cryptosystems based on word problems, *ICOMIDC Symp. Math. of Computation*, Ho Chi Minh City, April 1988.

Y. Driencourt, J. Michon (1987): Elliptic codes over a field of characteristic 2, *J. Pure Appl. Algebra* **45**, 15–39.

T. ElGamal (1985a): A public key cryptosystem and a signature scheme based on discrete logarithms, *IEEE Trans. Information Theory* **31**, 469–472.

T. ElGamal (1985b): A subexponential-time algorithm for computing discrete logarithms over $GF(p^2)$, *IEEE Trans. Information Theory* **31**, 473–481.

G. Faltings (1995): The proof of Fermat's Last Theorem by R. Taylor and A. Wiles, *Notices of the Amer. Math. Soc.* **42**, 743–746.

J. Feigenbaum, S. Kannan, N. Nisan (1990): Lower bounds on random-self-reducibility (extended abstract), *Fifth Annual Structure in Complexity Theory Conference*, IEEE Comput. Soc. Press, 100–109.

J. Feigenbaum, R. Lipton, S.R. Mahaney (1989): A completeness theorem for almost everywhere invulnerable generators, *Technical Memorandum*, AT&T Bell Laboratories.

M.R. Fellows, N. Koblitz (1993): Kid Krypto, *Advances in Cryptology – Crypto '92*, Springer-Verlag, 371–389.

M.R. Fellows, N. Koblitz (1994a): Combinatorially based cryptography for children (and adults), *Congressus Numerantium* **99**, 9–41.

M.R. Fellows, N. Koblitz (1994b): Combinatorial cryptosystems galore!, *Contemporary Math.* **168**, 51–61.

G. Frey, H. Rück (1994): A remark concerning m-divisibility and the discrete logarithm in the divisor class group of curves, *Math. Comp.* **62**, 865–874.

W. Fulton (1969): *Algebraic Curves*, Benjamin.

M.R. Garey, D.S. Johnson (1979): *Computers and Intractability: A Guide to the Theory of NP-Completeness*, W.H. Freeman & Co.

J. von zur Gathen (1990a): Functional decomposition of polynomials: the tame case, *J. Symbolic Comp.* **9**, 281–299.

J. von zur Gathen (1990b): Functional decomposition of polynomials: the wild case, *J. Symbolic Comp.* **10**, 437–452.

G. van der Geer (1991): Codes and elliptic curves, in *Effective Methods in Algebraic Geometry*, Birkhäuser, 159–168.

G. van der Geer, J. van Lint (1988): *Introduction to Coding Theory and Algebraic Geometry*, Birkhäuser.

M. Giusti (1984): Some effectivity problems in polynomial ideal theory, *EUROSAM 84: Proc. Intern. Symp. on Symbolic and Algebraic Computation, Cambridge, England*, Springer-Verlag, 159–171.

S. Goldwasser, J. Kilian (1986): Almost all primes can be quickly certified, *Proc. 18th ACM Symp. Theory of Computing*, 316–329.

S. Goldwasser, S. Micali (1982): Probabilistic encryption and how to play mental poker keeping secret all partial information, *Proc. 14th ACM Symp. Theory of Computing*, 365–377.

S. Goldwasser, S. Micali (1984): Probabilistic encryption, *J. Comput. System Sci.* **28**, 270–299.

S. Golomb (1982): *Shift Register Sequences*, 2nd ed., Aegean Park Press.

D.M. Gordon (1993): Discrete logarithms in $GF(p)$ using the number field sieve, *SIAM J. Discrete Math.* **6**, 124–138.

D.M. Gordon (1995): Discrete logarithms in $GF(p^n)$ using the number field sieve, preprint.

D.M. Gordon, K. McCurley (1993): Massively parallel computation of discrete logarithms, *Advances in Cryptology – Crypto '92*, Springer-Verlag, 312–323.

L. Goubin, J. Patarin (1998a): Trapdoor one-way permutations and multivariate polynomials, to appear.

L. Goubin, J. Patarin (1998b): A new analysis of Matsumoto-Imai like cryptosystems, to appear.

P. Guan (1987): Cellular automaton public-key cryptosystem, *Complex Systems* **1**, 51–56.

R. Gupta, M.R. Murty (1986): Primitive points on elliptic curves, *Compositio Math.* **58**, 13–44.

R.K. Guy (1981): *Unsolved Problems in Number Theory*, Springer-Verlag.

G. Harper, A. Menezes, S.A. Vanstone (1993): Public-key cryptosystems with very small key lengths, *Advances in Cryptology – Eurocrypt '92*, Springer-Verlag, 163–173.

M.E. Hellman, R.C. Merkle (1978): Hiding information and signatures in trapdoor knapsacks, *IEEE Trans. Information Theory* **24**, 525–530.

M.E. Hellman, S. Pohlig (1978): An improved algorithm for computing logarithms over $GF(p)$ and its cryptographic significance, *IEEE Trans. Information Theory* **24**, 106–110.

M.E. Hellman, J.M. Reyneri (1983): Fast computation of discrete logarithms in $GF(q)$, *Advances in Cryptology – Crypto '82*, Plenum Press, 3–13.

I.N. Herstein (1975): *Topics in Algebra*, 2nd ed., Wiley.

L.S. Hill (1931): Concerning certain linear transformation apparatus of cryptography, *Amer. Math. Monthly* **38**, 135–154.

M.-D. Huang, D. Ierardi (1994): Efficient algorithms for the effective Riemann–Roch problem and for addition in the Jacobian of a curve, *J. Symbolic Comp.* **18**, 519–539.

D. Husemöller (1987): *Elliptic Curves*, Springer-Verlag.

D.T. Huynh (1986a): A superexponential lower bound for Gröbner bases and Church-Rosser commutative Thue systems, *Information and Control* **68**, 196–206.

D.T. Huynh (1986b): The complexity of the membership problem for two subclasses of polynomial ideals, *SIAM J. Comput.* **15**, 581–594.

H. Imai, T. Matsumoto (1985): Algebraic methods for constructing asymmetric cryptosystems, *Algebraic Algorithms and Error-Correcting Codes, Proc. Third Intern. Conf., Grenoble, France*, Springer-Verlag, 108–119.

H. Imai, T. Matsumoto (1989): Public quadratic polynomial-tuples for efficient signature-verification and message-encryption, *Advances in Cryptology – Eurocrypt '88*, Springer-Verlag, 419–453.

R. Impagliazzo (1995): A personal view of average–case complexity, *IEEE Trans. Information Theory*, 134–147.

K. Ireland, M.I. Rosen (1990): *A Classical Introduction to Modern Number Theory*, 2nd ed., Springer-Verlag.

D.S. Johnson (1990): A catalog of complexity classes, in *Handbook of Theoretical Computer Science*, Vol. A, Elsevier, 67–161.

B. Kaliski (1987): A pseudorandom bit generator based on elliptic logarithms, *Advances in Cryptology – Crypto '86*, Springer-Verlag, 84–103.

B. Kaliski (1991): One-way permutations on elliptic curves, *J. Cryptology* **3**, 187–199.

J. Kari (1992): Cryptosystems based on reversible cellular automata, unpublished manuscript.

D.E. Knuth (1973): *The Art of Computer Programming. Vol. 3*, Addison-Wesley.

D.E. Knuth (1981): *The Art of Computer Programming. Vol. 2*, 2nd ed., Addison-Wesley.

K. Kobayashi, Y. Nemoto, K. Tamura (1990): Public key cryptosystem using multivariate polynomials, *Proc. Symp. on Cryptography and Information Security, Nihondaira, Japan*.

N. Koblitz (1987): Elliptic curve cryptosystems, *Math. Comp.* **48**, 203–209.

N. Koblitz (1988): Primality of the number of points on an elliptic curve over a finite field, *Pacific J. Math.* **131**, 157–165.

N. Koblitz (1989): Hyperelliptic cryptosystems, *J. Cryptology* **1**, 139–150.

N. Koblitz (1990): A family of jacobians suitable for discrete log cryptosystems, *Advances in Cryptology – Crypto '88*, Springer-Verlag, 94–99.

N. Koblitz (1991a): Constructing elliptic curve cryptosystems in characteristic 2, *Advances in Cryptology – Crypto '90*, Springer-Verlag, 156–167.

N. Koblitz (1991b): Elliptic curve implementation of zero-knowledge blobs, *J. Cryptology* **4**, 207–213.

N. Koblitz (1991c): Jacobi sums, irreducible zeta-polynomials, and cryptography, *Canadian Math. Bull.* **34**, 229–235.

N. Koblitz (1992): CM-curves with good cryptographic properties, *Advances in Cryptology – Crypto '91*, Springer-Verlag, 279–287.

N. Koblitz (1993): *Introduction to Elliptic Curves and Modular Forms*, 2nd ed., Springer-Verlag.

N. Koblitz (1994): *A Course in Number Theory and Cryptography*, 2nd ed., Springer-Verlag.

N. Koblitz (1997): Cryptography as a teaching tool, in *Cryptologia* **21**, 317–326.

N. Koblitz, A. Menezes, S.A. Vanstone (1998): The state of elliptic curve cryptography, to appear in *Designs, Codes and Cryptography*.

P. Kocher (1996): Timing attacks on implementations of Diffie–Hellman, RSA, DSS, and other systems, *Advances in Cryptology – Crypto '96*, Springer-Verlag, 104–113.

J. Kollár (1988): Sharp effective Nullstellensatz, *J. Amer. Math. Soc.* **1**, 963–975.

K. Koyama, U. Maurer, T. Okamoto, S.A. Vanstone (1993): New public-key schemes based on elliptic curves over the ring $\mathbb{Z}_n$, *Advances in Cryptology – Crypto '91*, Springer-Verlag, 252–266.

L. Kučera, S. Micali (1988): Cryptography and random graphs, unpublished manuscript.

B. LaMacchia, A.M. Odlyzko (1991): Computation of discrete logarithms in prime fields, *Designs, Codes and Cryptography* **1**, 47–62.

S. Lang (1978): *Cyclotomic Fields*, Springer-Verlag.

S. Lang (1984): *Algebra*, 2nd ed., Addison-Wesley.

G. Lay, H. Zimmer (1994): Constructing elliptic curves with given group order over large finite fields, *Algorithmic Number Theory*, Lect. Notes Comp. Sci. **877**, Springer-Verlag, 250–263.

F. Lehmann, M. Maurer, V. Müller, V. Shoup (1994): Counting the number of points on elliptic curves over finite fields of characteristic greater than three, *Algorithmic Number Theory*, Lect. Notes Comp. Sci. **877**, Springer-Verlag, 60–70.

A.K. Lenstra, H.W. Lenstra, Jr. (1993): *The Development of the Number Field Sieve*, Lect. Notes Math. **1554**, Springer-Verlag.

A.K. Lenstra, H.W. Lenstra, Jr., L. Lovász (1982): Factoring polynomials with rational coefficients, *Math. Ann.* **261**, 515–534.

H.W. Lenstra, Jr. (1975): Euclid's algorithm in cyclotomic fields, *J. London Math. Soc.* **10**, 457–465.

H.W. Lenstra, Jr. (1987): Factoring integers with elliptic curves, *Annals Math.* **126**, 649–673.

H.W. Lenstra, Jr., J. Pila, C. Pomerance (1993): A hyperelliptic smoothness test. I, *Philos. Trans. Roy. Soc. London* **345**, 397–408.

R. Lercier (1996): Computing isogenies in $\mathbb{F}_{2^n}$, in Henri Cohen, ed., *Algorithmic Number Theory, Proc. Second Intern. Symp., ANTS-II*, Springer-Verlag, 197–212.

R. Lercier, F. Morain (1995): Counting the number of points on elliptic curves over finite fields: strategies and performances, *Advances in Cryptology – Eurocrypt '95*, Springer-Verlag, 79–94.

R. Lercier, F. Morain (1996): Counting points on elliptic curves over $\mathbb{F}_{p^n}$ using Couveignes' algorithm, preprint.

L. Levin (1984): Problems complete in "average" instance, *Proc. 16th ACM Symp. Theory of Computing*, 465.

R. Lidl, H. Niederreiter (1986): *Introduction to Finite Fields and Their Applications*, Cambridge Univ. Press.

R. Majercik (1989): *The Chaldean Oracles: Text, Translation, and Commentary*, E.J. Brill.

U. Maurer, S. Wolf (1998): The security of the Diffie-Hellman protocol, to appear in *Designs, Codes and Crypography*.

E. Mayr, A. Meyer (1982): The complexity of the word problem for commutative semi-groups and polynomial ideals, *Advances in Math.* **46**, 305–329.

B. Mazur (1977): Modular curves and the Eisenstein ideal, *Inst. Hautes Études Sci. Publ. Math.* **47**, 33–186.

K. McCurley (1990a): The discrete logarithm problem, *Cryptology and Computational Number Theory, Proc. Symp. Appl. Math.* **42**, 49–74.

K. McCurley (1990b): Odds and ends from cryptology and computational number theory, *Cryptology and Computational Number Theory, Proc. Symp. Appl. Math.* **42** (1990), 145–166.

W. Meier, O. Staffelbach (1993): Efficient multiplication on certain non-supersingular elliptic curves, *Advances in Cryptology – Crypto '92*, Springer-Verlag, 333–344.

A. Menezes (1993): *Elliptic Curve Public Key Cryptosystems*, Kluwer Acad. Publ.

A. Menezes, T. Okamoto, S.A. Vanstone (1993): Reducing elliptic curve logarithms to logarithms in a finite field, *IEEE Trans. Information Theory* **39**, 1639–1646.

A. Menezes, P. van Oorschot, S.A. Vanstone (1996): *Handbook of Applied Cryptography*, CRC Press.

A. Menezes, S.A. Vanstone (1990): The implementation of elliptic curve cryptosystems, *Advances in Cryptology – Auscrypt '90*, Springer-Verlag, 2–13.

A. Menezes, S.A. Vanstone (1993): Elliptic curve cryptosystems and their implementation, *J. Cryptology* **6**, 209–224.

A. Menezes, S.A. Vanstone, R.J. Zuccherato (1993): Counting points on elliptic curves over $\mathbb{F}_{2^m}$, *Math. Comp.* **60**, 407–420.

V. Miller (1986): Uses of elliptic curves in cryptography, *Advances in Cryptology – Crypto '85*, Springer-Verlag, 417–426.

J.S. Milne (1986a): Abelian varieties, in G. Cornell and J.H. Silverman, eds., *Arithmetic Geometry*, Springer-Verlag, 103–150.

J.S. Milne (1986b): Jacobian varieties, in G. Cornell and J.H. Silverman, eds., *Arithmetic Geometry*, Springer-Verlag, 167–212.

H.M. Möller, F. Mora (1984): Upper and lower bounds for the degree of Gröbner bases, *EUROSAM 84: Proc. Intern. Symp. on Symbolic and Algebraic Computation, Cambridge, England*, Springer-Verlag, 172–183.

T. Mora *et al.* [pseudonyms Boo Barkee, Deh Cac Can, Julia Ecks, Theo Moriarty, R.F. Ree] (1993): Why you cannot even hope to use Gröbner bases in public key cryptography: An open letter to a scientist who failed and a challenge to those who have not yet failed, unpublished manuscript.

F. Morain (1991): Building cyclic elliptic curves modulo large primes, *Advances in Cryptology – Eurocrypt '91*, Springer-Verlag, 328–336.

L.J. Mordell (1922): On the rational solutions of the indeterminate equations of the third and fourth degrees, *Proc. Camb. Phil. Soc.* **21**, 179–192.

V. Müller, S.A. Vanstone, R. Zuccherato (1998): Discrete logarithm based cryptosystems in quadratic function fields of characteristic 2, *Designs, Codes and Cryptography* **14**, 159–178.

R. Mullin, I. Onyszchuk, S.A. Vanstone, R. Wilson (1988/1989): Optimal normal bases in $GF(p^n)$, *Discrete Appl. Math.* **22**, 149–161.

D. Mumford (1984): *Tata Lectures on Theta II*, Birkhäuser.

P.S. Novikov (1955): On the algorithmic unsolvability of the word problem in group theory, *Trudy Mat. Inst. im. Steklova* **44**, 1–143.

A.M. Odlyzko (1985): Discrete logarithms in finite fields and their cryptographic significance, *Advances in Cryptology – Eurocrypt '84*, Springer-Verlag, 224–314.

A.M. Odlyzko (1990): The rise and fall of knapsack cryptosystems, *Cryptology and Computational Number Theory, Proc. Symp. Appl. Math.* **42**, 75–88.

A.M. Odlyzko (1995): The future of integer factorization, *CryptoBytes* **1**, No. 2, 5–12.

S. O'Malley, H. Orman, R. Schroeppel, O. Spatscheck (1995): Fast key exchange with elliptic curve systems, *Advances in Cryptology – Crypto '95*, Springer-Verlag, 43–56.

C.H. Papadimitriou (1994): *Computational Complexity*, Addison-Wesley.

J. Patarin (1995): Cryptanalysis of the Matsumoto and Imai public key scheme of Eurocrypt '88, *Advances in Cryptology – Crypto '95*, Springer-Verlag, 248–261.

J. Patarin (1996a): Hidden fields equations (HFE) and isomorphisms of polynomials (IP): two new families of asymmetric algorithms, *Advances in Cryptology – Eurocrypt '96*, Springer-Verlag, 33–48.

J. Patarin (1996b): Asymmetric cryptography with a hidden monomial, *Advances in Cryptology – Crypto '96*, Springer-Verlag, 45–60.

M. Petersen (1994): Hyperelliptic cryptosystems, *Technical Report*, Univ. Aarhus, Denmark.

J. Pila (1990): Frobenius maps of abelian varieties and finding roots of unity in finite fields, *Math. Comp.* **55**, 745–763.

J. Pollard (1978): Monte Carlo methods for index computation mod p, *Math. Comp.* **32**, 918–924.

B. Poonen (1996): Computational aspects of curves of genus at least 2, in Henri Cohen, ed., *Algorithmic Number Theory, Proc. Second Intern. Symp., ANTS-II*, Springer-Verlag, 283–306.

G. Purdy (1974): A high-security log-in procedure, *Communications of the ACM* **17**, 442–445.

M.O. Rabin (1980): Probabilistic algorithms for testing primality, *J. Number Theory* **12**, 128–138.

K. Ribet (1990): On modular representations of Gal($\overline{\mathbb{Q}}, \mathbb{Q}$) arising from modular forms, *Invent. Math.* **100**, 431–476.

R. Rivest (1990): Cryptography, in *Handbook of Theoretical Computer Science*, Vol. A, Elsevier, 717–755.

R. Rivest, A. Shamir, L.N. Adleman (1978): A method for obtaining digital signatures and public-key cryptosystems, *Communications of the ACM* **21**, 120–126.

H.E. Rose (1994): *A Course in Number Theory*, 2nd ed., Clarendon Press.

K. Rosen (1993): *Elementary Number Theory and Its Applications*, 3rd ed., Addison-Wesley.

R. Scheidler, A. Stein, H.C. Williams (1996): Key-exchange in real quadratic congruence function fields, *Designs, Codes and Cryptography* **7**, 153–174.

R. Scheidler, H.C. Williams (1995): A public-key cryptosystem utilizing cyclotomic fields, *Designs, Codes and Cryptography* **6**, 117–131.

O. Schirokauer (1993): Discrete logarithms and local units, *Philos. Trans. Roy. Soc. London* **345**, 409–423.

C.P. Schnorr (1991): Efficient signature generation by smart cards, *J. Cryptology* **4**, 161–174.

R. Schoof (1985): Elliptic curves over finite fields and the computation of square roots mod p, *Math. Comp.* **44**, 483–494.

R. Schoof (1987): Nonsingular plane cubic curves, *J. Combinatorial Theory, Ser. A* **46**, 183–211.

E. Seah, H.C. Williams (1979): Some primes of the form $(a^n - 1)/(a - 1)$, *Math. Comp.* **33**, 1337–1342.

A.L. Selman (1988): Complexity issues in cryptography, *Computational Complexity Theory, Proc. Symp. Appl. Math.* **38**, 92–107.

A. Shamir (1984): A polynomial time algorithm for breaking the basic Merkle–Hellman cryptosystem, *IEEE Trans. Information Theory* **30**, 699–704.

A. Shamir (1992): IP=PSPACE, *Journal of the ACM* **39**, 869–877.

D. Shanks (1972): Five number-theoretic algorithms, *Congressus Numerantium* **7**, 51–70.

D. Shanks (1985): *Solved and Unsolved Problems in Number Theory*, 3rd ed., Chelsea.

C.E. Shannon (1949): Communication theory of secrecy systems, *Bell Syst. Tech. J.* **28**, 656–715.

J. Silverman (1986): *The Arithmetic of Elliptic Curves*, Springer-Verlag.

J. Silverman (1994): *Advanced Topics in the Arithmetic of Elliptic Curves*, Springer-Verlag.

J. Solinas (1997): An improved algorithm for arithmetic on a family of elliptic curves, *Advances in Cryptology – Crypto '97*, Springer-Verlag, 357–371.

J. Tate (1965): Algebraic cycles and poles of zeta functions, *Proc. Purdue Conf., 1963*, 93–110.

R. Taylor, A. Wiles (1995): Ring-theoretic properties of certain Hecke algebras, *Annals Math.* **141**, 553–572.

J. Tunnell (1983): A classical Diophantine problem and modular functions of weight 3/2, *Invent. Math.* **72**, 323–334.

P. van Oorschot (1992): A comparison of practical public-key cryptosystems based on integer factorization and discrete logarithms, in G. Simmons, ed., *Contemporary Cryptology: The Science of Information Integrity*, IEEE Press, 289–322.

P. van Oorschot, M. Wiener (1994): Parallel collision search with application to hash functions and discrete logarithms, Proc. 2nd ACM Conf. on Computer and Communications Security, Fairfax, Virginia, 210–218.

P. van Oorschot, M. Wiener (1998): Parallel collision search with cryptanalytic applications, to appear in *J. Cryptology*.

E. Volcheck (1994): Computing in the Jacobian of a plane algebraic curve, *Algorithmic Number Theory*, Lect. Notes Comp. Sci. **877**, Springer-Verlag, 221–233.

W. Waterhouse (1969): Abelian varieties over finite fields, *Ann. Sci. École Norm. Sup.* **2**, 521–560.

D. Weber (1996): Computing discrete logarithms with the general number field sieve, in Henri Cohen, ed., *Algorithmic Number Theory, Proc. Second Intern. Symp., ANTS-II*, Springer-Verlag, 391–403.

A. Weil (1949): Numbers of solutions of equations in finite fields, *Bull. Amer. Math. Soc.* **55**, 497–508.

E.P. Wigner (1960): The unreasonable effectiveness of mathematics in the natural sciences, *Comm. Pure Appl. Math.* **13**, 1–14.

A. Wiles (1995): Modular elliptic curves and Fermat's Last Theorem, *Annals Math.* **141**, 443–551.

H.S. Wilf (1984): Backtrack: An $O(1)$ expected time graph coloring algorithm, *Inform. Process Lett.* **18**, 119–122.

M.V. Wilkes (1968): *Time-Sharing Computer Systems*, Elsevier.

S. Wolfram (1986): Cryptography with cellular automata, *Advances in Cryptology – Crypto '85*, Springer-Verlag, 429–432.

Subject Index